AF556221

FINITE MATHEMATICS

By

A.K. Sharma

DISCOVERY PUBLISHING HOUSE PVT. LTD.
NEW DELHI-110 002

First Published – 2009

Reprinted – 2017

ISBN: 978-81-8356-453-3

Finite Mathematics

Published by:

DISCOVERY PUBLISHING HOUSE PVT. LTD.

4383/4B, Ansari Road Darya Ganj
New Delhi - 110 002 (India)
Phone: +91-11-23279245, 43596064-65
Fax: +91-11-23253475
E-mail: discoverypublishinghouse@gmail.com
sales@discoverypublishinggroup.com
web: www.discoverypublishinggroup.com

Printed at:
Infinity Imaging Systems
Delhi

Preface

Finite Mathematics is a course, conventionally required for business students, in which the curriculum brings together several mathematical topics, including basic probability theory, with introduction to linear programming, some theory of matrices and determinates and sometimes abbreviated accounts of calculus. In other words, Finite mathematics is the branch of Mathematics, which does not involve infinite sets, limits or continuity.

Finite Mathematics is a very broad and heterogeneous area of mathematics, studying finite sets and configurations, with certain properties, their characterisation and in number. It is related to almost all other areas of mathematics and it also has a wide range of applications.

Modern high-speed computers have made all forms of computation vastly in an easy form. In past, while the calculations used to take several days to complete, these can now be carried out in a few seconds, with the help of computers. The availability of modern computers can take a great deal of dudgery, out of mathematical computations and makes possible large-scale computations that would have been impossible, otherwise.

Uptill now, a number of books on the subject have provided a certain level of knowledge. So we have made this humble effort to enhance the understanding of the subject by concerned readers. Expectedly, this effort would prove to be beneficial for the students and general readers, alike.

Enlightening and positive comments from the esteemed readers would be heartily appreciated.

Content

5+13

Sets and Counting Problems

Introduction

A well-defined collection of objects is known as a *set*. This concept, in its complete generality, is of great importance in mathematics since all of mathematics can be developed by starting from it.

The various pieces of furniture in a given room form a set. So do the books in a given library, or the integers between 1 and 1,000,000 or all the ideas that mankind has had, or the human beings alive between 1 billion BC and AD 10 billion. These examples are all examples of *finite* sets, that is, sets having a finite number of elements. All the sets discussed in this book will be finite sets.

The collection of all tall people is *not* a well-defined set, because the word "tall" is not precisely defined. On the other hand, the set of all people whose height is six feet or more *is* a well-defined set, because we can determine whether any given person belongs to the set simply by measuring his height.

There are two essentially different ways of specifying a set. One can give a rule by which it can be determined whether or not a given object is a member of the set, or one can give a complete list of the elements in the set. We shall say that the former is a *description* of the set and the latter is a *listing* of the set. For example, we can define a set of four people as:

(a) The members of the string quartet which played in town last night, or;

(b) Four particular persons whose names are Jones, Smith, Brown, and Green. It is customary to use braces to surround the listing of a set; thus, the set above should be listed (Jones, Smith, Brown, Green).

We shall frequently be interested in sets of logical possibilities, since the analysis of such sets is very often a major task in the solving of a problem. Suppose, for example, that we were interested in the successes of three candidates who enter the presidential primaries (we assume there are no other entries). Suppose that the key primaries will be held in New Hampshire, Minnesota, Winsonsin, and California. Assume that candidate A enters all the primaries, that B does not contest in New Hampshire's primary, and C does not contest in Wisconsin's. A list of the logical possibilities is given in Figure 1.1. Since the New Hampshire and Wisconsin primaries can each end in two ways, and the Minnesota and California primaries can each end in three ways, there are in all $2 \cdot 2 \cdot 3 \cdot 3 = 36$ different logical possibilities as listed in Figure 1.1.

Possibility Number	*Winner in New Hampshire*	*Winner in Minnesota*	*Winner in Wisconsin*	*Winner in California*
P1	A	A	A	A
P2	A	A	A	B
P3	A	A	A	C
P4	A	A	B	A
P5	A	A	B	B
P6	A	A	B	C
P7	A	B	A	A
P8	A	B	A	B
P9	A	B	A	C
P10	A	B	B	A
P11	A	B	B	B
P12	A	B	B	C
P13	A	C	A	A

P14	A	C	A	B
P15	A	C	A	C
P16	A	C	B	A
P17	A	C	B	B
P18	A	C	B	C
P19	C	A	A	A
P20	C	A	A	B
P21	C	A	A	C
P22	C	A	B	A
P23	C	A	B	B
P24	C	A	B	C
P25	C	B	A	A
P26	C	B	A	B
P27	C	B	A	C
P28	C	B	B	A
P29	C	B	B	B
P30	C	B	B	C
P31	C	C	A	A
P32	C	C	A	B
P33	C	C	A	C
P34	C	C	B	A
P35	C	C	B	B
P36	C	C	B	C

Figure: 1.1

A set that consists of some members of another set is called a *subset* of that set. For example, the set of those logical possibilities in Figure 1.1 for which the statement "Candidate A wins at least three primaries" is true, is a subset of the set of all logical possibilities. This subset can also be defined by listing its members: {P1, P2, P3, P4, P7, P13, P19}.

In order to discuss all the subsets of a given set, let us introduce the following terminology. We shall call the original set the *universal set*, one-element subsets will be called *unit sets*, and the set which contains no members the *empty set*. We do not introduce special names for other kinds of subsets of the universal set. As an example, let the universal set U consist of the three elements $\{a, b, c\}$. The *proper subsets* of U are those sets containing some but not all of the elements of U. The proper subsets here consist of three two-element sets—namely, $\{a, b\}$, $\{a, c\}$, and $\{b, c\}$—and three unit sets—namely, $\{a\}$, $\{b\}$, and $\{c\}$. To complete the picture, we also consider the universal set a subset (but not a proper subset) of itself, and we consider the empty set* E, which contains no elements of U, as a subset of U. At first it may seem strange that we should include the sets U and E as subsets of U, but the reasons for their inclusion will become clear later.

We saw that the three-element set above had $8 = 2^3$ subsets. In general, a set with n elements has 2^n subsets, as can be seen in the following manner. We form subsets P of U by considering each of the elements of U in turn and deciding whether or not to include it in the subset P. If we decide to put every element of U into P, we get the universal set, and if we decide to put no element of U into P, we get the empty set. In most cases, we shall put some but not all the elements into P and thus obtain a proper subset of U. We have to make n decisions, one for each element of the set, and for each decision, we have to choose between two alternatives. We can make these decisions in $2 \cdot 2 \cdot \ldots \cdot 2 = 2^n$ ways, and hence this is the number of different subsets of U that can be formed. Observe that our formula would not have been so simple if we had not included the universal set and the empty set as subsets of U.

In the example of the voting primaries above, there are 2^{36} or about 70 billion subsets. Of course, we cannot deal with these many subsets in a practical problem, but fortunately we are usually interested in only a few of the subsets. The most interesting subsets are those which can be defined by means of a simple rule such as "the set of all logical possibilities in which C loses at least two primaries." It would be difficult to give a simple description for the subset containing the elements {P1, P4, P14, P30, P34}. On the other hand, we shall see in the next section how to define new subsets in terms of subsets already defined.

Examples: We illustrate the two different ways of specifying sets in terms of the primary voting example. Let the universal set U be the logical possibilities given in Figure 1.1.

1. What is the subset of U in which candidate B wins more primaries than either of the other candidates?

   Answer: (P11, P12, P17, P23, P26, P28, P29).

2. What is the subset in which the primaries are split two and two?

   Answer: (P5, P8, P10, P15, P21, P30, P31, P35).

3. Describe the set (P1, P4, P19, P22).

   Answer: The set of possibilities for which A wins in Minnesota and California.

4. How can we describe the set {P18, P24, P27}?

   Answer: The set of possibilities for which C wins in California, and the other primaries are split three ways.

Exercises

1. In the primary example, list each of the following sets.
   (a) The set in which A and C win the same number of primaries.
   (b) The set in which the winner of the New Hampshire primary does not win another primary.
   (c) The set in which C wins all four primaries.
2. Again referring to the primary example, give simple descriptions of the following sets.
   (a) [P1, P4, P8, P11, P15, P18, P19, P22, P26, P29, P33, P36]
   (b) [P18, P22, P26].
   (c) [P1, P11, P19, P29].
3. The primaries are considered decisive if a candidate can win thre primaries, or if he wins two primaries including California. List th set in which the primaries are decisive.
4. List the set of four-letter "words" formed by writing down th letters of the word *stop* in all possible ways. *[Hint:* The set ha 24 elements.]

5. In Exercise 4, list the following subsets:
   (a) The set of English words.

   [Partial Ans. There are 6]
   (b) The set in which the letters are in alphabetical order either from left to right or from right to left.
   (c) The set in which p and t are next to each other.
   (d) The set in which only s is between o and t.
   (e) The set in which t and s are at the ends.
6. Find all pairs in Exercise 5 in which one set is a subset of the other.
7. A baker has four feet of display space to fill with some combination of bread, cake, and pie. A loaf of bread takes one-half foot of space, a cake takes one foot, and a pie takes two feet. Construct the set of possible distributions of shelf space, considering only the *total* space allotted to each kind of item.
8. In Exercise 7, list the following subsets:
   (a) The set in which as much space is devoted to pie as to cake.
   (b) The set in which equal space is given to two different items, and at least two different items are displayed.
   (c) The set in which six or more items are displayed.
   (d) The set in which at least two of the above conditions are satisfied.
9. A man has 65 cents in change, but he has no pennies and has at least as many dimes as nickels. Find the set of possibilities for his collection of coins.
10. In Exercise 9, list the following subsets:
    (a) The set in which the man has exactly one quarter.
    (b) The set in which the man has more half-dollars than quarters.
    (c) The set in which the man has fewer than six coins.
    (d) The set in which none of the above conditions is satisfied.
11. A set has 51 elements. How many subsets does it have? How many of the subsets have an even number of elements?

    [*Ans.* 2^{51}, 2^{50}]
12. Do Exercise 11 for the case of a set with 52 elements.

Operations on Subsets

Earlier, we considered the ways in which one could form new statements from given statements. Now, we shall consider an analogous procedure, the formation of new sets from given sets. We shall assume that each of the sets that we use in the combination is a subset of some universal set, and we shall also want the newly-formed set to be a subset of the same universal set. As usual, we can specify a newly-formed set either by a description or by a listing.

If P and Q are two sets, we shall define a new set $P \cap Q$, called the *intersection* of P and Q as follows: $P \cap Q$ is the set which contains those and only those elements which belong to both P and Q. As an example, consider the logical possibilities listed in Figure 1.1. Let P be the subset in which candidate A wins at least three primaries, i.e., the set {P1, P2, P3, P4, P7, P13, P19}; let Q be the subset in which A wins the first two primaries, i.e., the set {P1, P2, P3, P4, P5, P6}. Then the intersection $P \cap Q$ is the set in which both events take place, i.e., where A wins the first two primaries *and* wins at least three primaries. Thus, $P \cap Q$ is the set {P1, P2, P3, P4}.

If P and Q are two sets, we shall define a new set $P \cup Q$ called the *union* of P and Q as follows: $P \cup Q$ is the set that contains those and only those elements that belong either to P or to Q (or to both). In the example in the paragraph above, the union $P \cup Q$ is the set of possibilities for which either A wins the first two primaries *or* wins at least three primaries, i.e., the set {P1, P2, P3, P4, P5, P6, P7, P13, P19}.

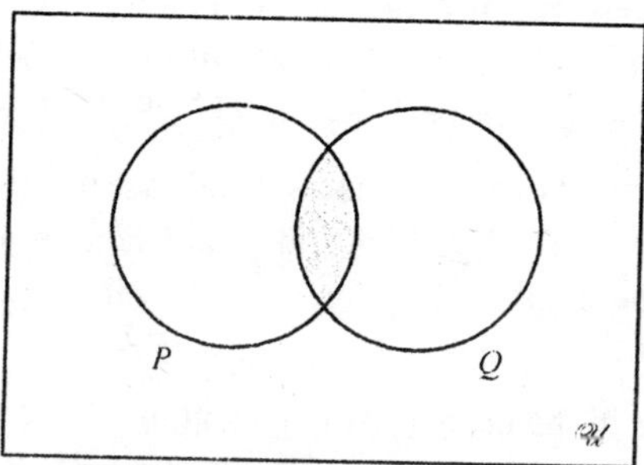

Figure: 1.2

To help in visualising these operations, we shall draw diagrams, called *Venn diagrams,* which illustrate them. We let the universal set be a rectangle and let Subsets be circles drawn inside the rectangle. In Figure 1.2

we show two sets P and Q as shaded circles, P shaded in colour and Q in gray. Then the area shaded in both colour and gray is the intersection Pflg and the total shaded area is the union $P \cup Q$.

If P is a given subset of the universal set U, we can define a new set $\tilde{P}$ called the *complement* of P as follows: $\tilde{P}$ is the set of all elements of 11 that are *not* contained in P. For example, if, as above, Q is the set in which candidate A wins the first two primaries, then $\tilde{Q}$ is the set {P7, P8,. . ., P36}. The shaded area in Figure 1.3 is the complement of the set P. Observe that the complement of the empty set E is the universal set U, and also that the complement of the universal set is the empty set.

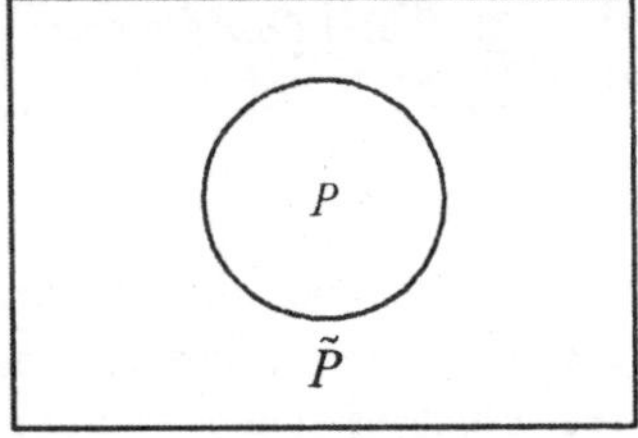

Figure: 1.3

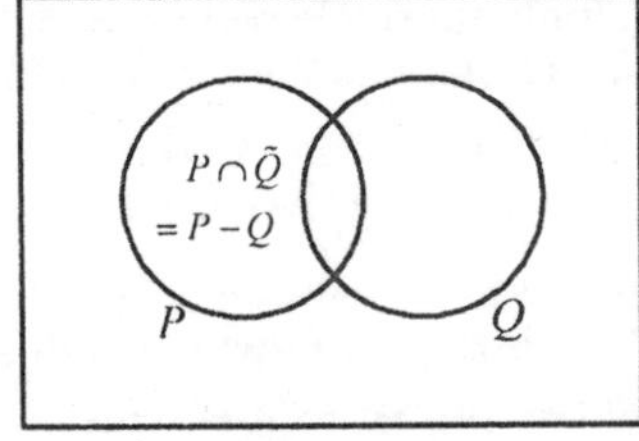

Figure: 1.4

Sometimes we shall be interested in only part of the complement of a set. For example, we might wish to consider the part of the complement of the set Q that is contained in P, i.e., the set $P \cap \tilde{Q}$. The shaded area in Figure 1.4 is $P \cap \tilde{Q}$.

A somewhat more suggestive definition of this set can be given as follows: Let $P - Q$ be the *difference* of P and Q, that is, the set that contains those elements of P that do not belong to Q. Figure 1.4 shows that $P \cap \tilde{Q}$ and $P - Q$ are the same set. In the primary voting example above, the set $P - Q$ can be listed as {P7, P13, P19}.

The complement of a subset is a special case of a difference set, since we can write $\tilde{Q} = U - Q$. If P and Q are non-empty Subsets whose intersection is the empty set, i.e., $P \cap Q = E$, then we say that they are *disjoint* Subsets.

Example 1: In the primary voting example, let R be the set in which A wins the first three primaries, i.e., the set {P1, P2, P3}; let S be the set in which A wins the last two primaries, i.e., the set {P1, P7, P13, P19, P25, P31}. Then, $R \cap S$ = {P1} is the set in which A wins the first three primaries and also the last two, that is, he wins all the primaries. We also have

$$R \cup S = \{P1, P2, P3, P7, P13, P19, P25, P31\},$$

which can be described as the set in which A wins the first three primaries or the last two. The set in which A does not win the first three primaries is $\tilde{R}$ = {P4, P5, . . ., P36}. Finally, we see that the difference set $R - S$ is the set in which A wins the first three primaries but not both of the last two. This set can be found by taking from R the element PI which it has in common with S, so that $R - S$ = {P2, P3}.

Example 2: Let us give a step-by-step construction of the Venn diagram for the set $(P \cap Q) \cup (\tilde{P} \cap \tilde{Q})$. Figure 1.5 shows the set $P \cap Q$ which is the same as the set of Figure 1.2 shaded in both colour and gray; Figure 1.6 shows the set $\tilde{P} \cap \tilde{Q}$ which is the same as the complement of the shaded area in Figure 1.2. Finally, Figure 1.7 is the union of the two areas in Figures 1.5 and 1.6 and is the answer desired.

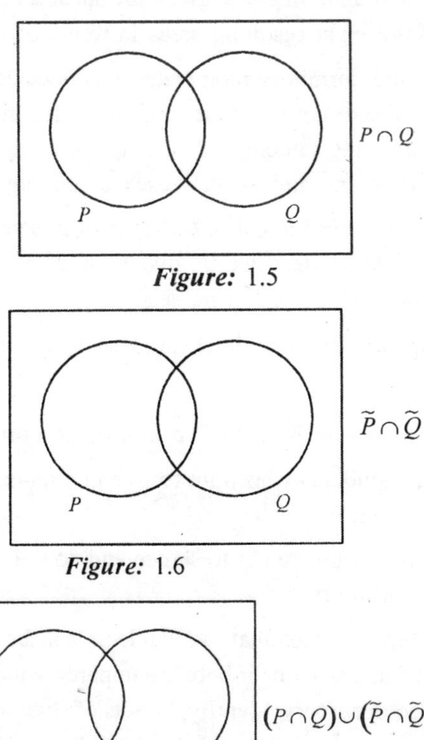

Figure: 1.5

Figure: 1.6

Figure: 1.7

Exercises

1. Draw Venn diagrams for the following sets:
   (a) $P \cap Q$.
   (b) $\tilde{P} \cup Q$.
   (c) $P \cup \tilde{Q}$.
   (d) $\tilde{P} \cup \tilde{Q}$.
2. Give a step-by-step construction of the diagram for $((P \cup Q) - (P \cap Q)) \cap \tilde{Q}$.
3. Venn diagrams are also useful when three subsets are given. Construct such a diagram, given the subsets P, Q, and R. Identify each of the eight resulting areas in terms of P, Q, and R.
4. In assigning dormitory roommates, a college considers a student's sex, whether or not the student wants to live in a co-ed dormitory, and whether the student is a freshman or an upperclassman. Draw a Venn diagram, and identify each of the eight areas.
5. Let F be the set of females, U the set of upperclassmen, and C the set of students desiring to live in a co-ed dormitory. Define (symbolically) the following sets:
   (a) Upperclass males who do not want to live in a co-ed dormitory.
   [*Ans.* $U \cap \tilde{F} \cap \tilde{C}$]
   (b) Women who want to live in a co-ed dormitory.
   (c) Male students who want to live in a co-ed dormitory and are freshmen.
   (d) Women who are not freshmen and do not want to live in a co-ed dormitory.
6. The college decides that two students can be roommates if both are of the same sex or if both are upperclassmen who want to live in a co-ed dormitory. Identify the sets of students with the property that any two members of the set can be roommates.
7. The results of a survey of church attendance and golf playing are given in the following table:

Occupation	Golfs and Attends	Golfs and Doesn't Attend	Doesn't Golfs and Attends	Doesn't Golf and Doesn't Attend
Doctor	15	20	3	2
Lawyer	10	9	9	6
CPA	8	0	11	7

Let D = doctor, L = lawyer, C = CPA, G = golfs, A = attends. Determine the number of people in each of the following classes.

(a) $D \cap G \cap \tilde{A}$.

(b) $\tilde{C} \cap \tilde{G} \cap A$.

(c) $\left(\widetilde{G \cup A}\right) \cap L$..

(d) $(D \cup L) \cap G$. [Ans. 54]

(e) $\tilde{L} \cap \left((A \cap G) \cup (\tilde{A} \cap G)\right)$. [*Ans.* 43]

8. In Exercise 7, which set of each of the following pairs has more members?

(a) $(D \cap G) - A$ or $\tilde{L} \cup (G \cup A)$?

(b) E or $C \cap \tilde{A} \cap$ G?

(c) $\left(\widetilde{D \cup L}\right)$ or C?

9. A college student hired to survey 1000 beer drinkers and record their age, sex, and educational level turned in the following figures: 700 males, 600 people over 25 years of age, 400 college graduates, 250 male college graduates, 225 college graduates over 25, 350 males over 25, and 150 male college graduates over 25. After turning in his results, he was fired. Why? *[Hint:* Draw a Venn diagram with three circles—for males, college graduates, and those over 25. Fill in the numbers in each of the eight areas, using the data given above. Start from the end of the list and work back.]

10. A survey of 110 lung cancer patients showed that 70 were cigarette smokers, 60 lived in urban areas, and 35 had hazardous occupations. Forty of the smokers lived in urban areas, 15 had hazardous occupations, and 5 were in both categories. Ten of the patients with hazardous occupations neither lived in an urban area nor smoked.

(a) How many of the patients living in urban areas had hazardous occupations? [*Ans.* 15]

(b) How many of those living in the urban areas neither smoked nor had hazardous occupations? [*Ans.* 10]

(c) How many patients smoke if and only if they live in an urban area?

(d) How many patients neither smoked, nor lived in an urban area, nor had a hazardous occupation?

11. A second survey of 100 patients had the following results: 45 smokers who lived in urban areas, 37 of whom did not have a hazardous occupation; 20 people with hazardous occupations, of whom 10 live in urban areas and 10 smoke; 75 smokers; and 10 who neither smoke, nor have a hazardous occupation, nor live in an urban area.

(a) How many patients with hazardous occupations neither smoke nor live in an urban area? [*Ans.* 8]

(b) How many patients live in an urban area?

(c) How many patients smoke if and only if they do not have a hazardous occupation?

(d) How many patients smoke, have a hazardous occupation, and live in an urban area?

12. The following table summarises the responses of 100 students asked what they thought about during math lectures:

Class and Status	*Neither Food Nor Football*	*Only Food*	*Only Football*	*Food and Football*
Senior Majors	20	12	4	6
Senior Non-majors	8	10	15	0
Junior Majors	2	1	6	1
Junior Non-majors	3	5	5	2

All the categories can be defined in terms of the following four: *M* (majors), *S* (seniors), *F* (food), and *FT* (football). How many students fall into each of the following categories?

(a) S

(b) $S - M$

(f) $\tilde{J} \cup \widetilde{F}$ [*Ans.* 91]

(g) $S \cap \widetilde{M} \cap F$

(c) $M - S$

(d) $J \cap \widetilde{M} \cap F \cap FT$

(e) $(J \cap F)$

(h) $(S \cup F) - \widetilde{FT}$ [*Ans.* 28]

(i) $S \cap M \cap \left(\widetilde{F \cup FT}\right)$ [*Ans.* 20]

(j) $S \cup J$

Sets and Compound Statements

The reader may have observed several times in the preceding sections that there was a close connection between sets and statements, and between set operations and compounding operations. In this section, we shall formalise these relationships.

If we have a number of statements relative to a set of logical possibilities, there is a natural way of assigning a set to each statement. First, we take the set of logical possibilities as our universal set. Then to each statement we assign the subset of logical possibilities of the universal set for which that statement is true. This idea is so important that we embody it in a formal definition.

Definition: Let U be a set of logical possibilities, let p be a statement relative to it, and let P be that subset of the possibilities for which p is true; then we call P the *truth set of p.*

If p and q are statements, then $p \vee q$ and $p \wedge q$ are also statements and hence must have truth sets. To find the truth set of $p \vee q$, we observe that it is true whenever p is true or q is true (or both). Therefore, we must assign to $p \vee q$ the logical possibilities which are in P or in Q (or both); that is, we must assign to $p \vee q$ the set $P \cup Q$. On the other hand, the statement $p \wedge q$ is true only when both p and q are true, so that we must assign to $p \wedge q$ the set $P \cap Q$.

Thus, we see that there is a close connection between the logical operation of disjunction and the set operation of union, and also between conjunction and intersection. A careful examination of the definitions of union and intersection shows that the word "or" occurs in the definition of union and the word "and" occurs in the definition of intersection. Thus, the connection between the two theories is not surprising.

Since the connective "not" occurs in the definition of the complement of a set, it is not surprising that the truth set of $\sim p$ is $\widetilde{p}$. This follows since $\sim p$ is true when p is false, so that the truth set of $\sim p$ contains all logical possibilities for which p is false, that is, the truth set of $\sim p$ is $\widetilde{p}$.

The truth sets of two propositions p and q are shown in Figure 1.8. Also marked on the diagram are the various logical possibilities for these two statements. The reader should pick out in this diagram the truth sets of the statements $p \vee q$, $p \wedge q$, $\tilde{p}$ and $\sim q$.

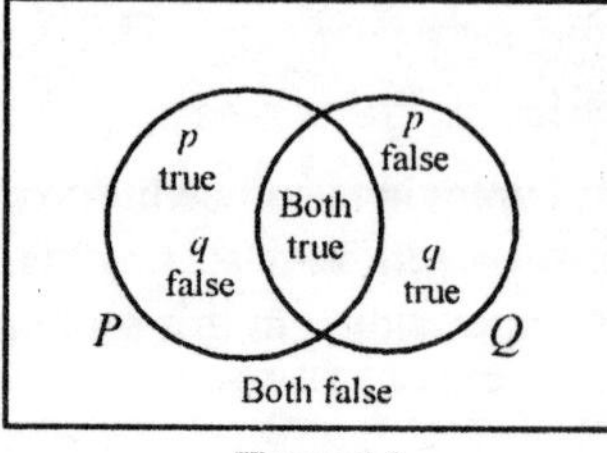

Figure: 1.8

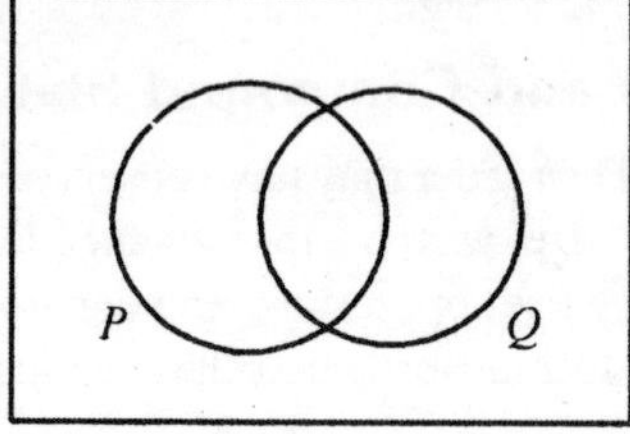

Figure: 1.9

The connection between a statement and its truth set makes it possible to "translate" a problem about compound statements into a problem about sets. It is also possible to go in the reverse direction. Given a problem about sets, think of the universal set as being a set of logical possibilities and think of a subset as being the truth set of a statement. Hence, we can "translate" a problem about sets into a problem about compound statements.

So far we have discussed only the truth sets assigned to compound statements involving $\vee$, $\wedge$, and $\sim$. All the other connectives can be defined in terms of these three basic ones, so that we can deduce what truth sets should be assigned to them. For example, we know that $p \rightarrow q$ is equivalent to $\sim p \vee q$. Hence, the truth set of $p \rightarrow q$ is the same as the truth set of $\sim p \vee q$, that is, it is $\tilde{P} \cup Q$. The Venn diagram for $p \rightarrow q$ is shown in Figure 1.9, where the shaded area is the truth set for the statement. Observe that the unshaded area in Figure 1.9 is the set $P - Q = P \cap \tilde{Q}$, which is the truth set of the statement $p \wedge \sim q$. Thus, the shaded area is the set $\widetilde{\left(P - Q\right)} = \widetilde{P \cap \tilde{Q}}$, which is the truth set of the statement $\sim [p \wedge \sim q]$. We have thus discovered the fact that $(p \rightarrow q)$, $(\sim p \vee q)$, and $(\sim p \wedge \sim q)$ are equivalent. It is always the case that two compound statements are equivalent if and only if they have the same truth sets. Thus, we can test for equivalence by checking whether they have the same Venn diagram.

Suppose that p is a statement that is logically true. What is its truth set? Now p is logically true if and only if it is true in every logically possible case, so that the truth set of p must be U. Similarly, if p is logically false, then it is false for every logically possible case, so that its truth set is the empty set E.

Finally, let us consider the implication relation. Recall that p implies q if and only if the conditional $p \rightarrow q$ is logically true. But $p \rightarrow q$ is logically true if and only if its truth set is U, that is, $\left(\widetilde{P-Q}\right) = U$, or $(P - Q) = E$. From Figure 1.4 we see that if $P - Q$ is empty, then P is contained in Q. We shall symbolise the containing relation as follows: $P \subset Q$ means "P is a subset of Q." We conclude that $p \Rightarrow q$ if and only if $P \subset Q$.

Figure 1.10 supplies a "dictionary" for translating from statement language to set language, and back. To each statement relative to a set of possibilities U there corresponds a subset of U—namely, the truth set of the statement.

Statement Language	Set Language
r	R
s	S
$\sim r$	$\widetilde{R}$
$r \vee s$	$R \cup S$
$r \wedge s$	$R \cap S$
$r \rightarrow s$	$\left(\widetilde{R-S}\right)$
$r \Rightarrow s$	$R \subset S$
$r \Leftrightarrow s$	$R = S$

Figure: 1.10

This is shown in lines 1 and 2 of the figure. To each connective, there corresponds an operation on sets, as illustrated in the next four lines. And to each relation between statements there corresponds a relation between sets, examples of which are shown in the last two lines of the figure.

Example 1: Verify by means of a Venn diagram that the statement $[p \vee (\sim p \vee q)]$ is logically true. The assigned set of this statement is $[P \cup (\widetilde{p} \cup Q)]$, and its Venn diagram is shown in Figure 1.11. In that figure, the set P is shaded in colour, and the set $\tilde{P} \cup Q$ is shaded in gray. Their union is the entire shaded area, which is U, so that the compound statement is logically true.

Example 2: Demonstrate by means of Venn diagrams that $p \vee (q \wedge r)$ is equivalent to $(p \vee q) \wedge (p \vee r)$.

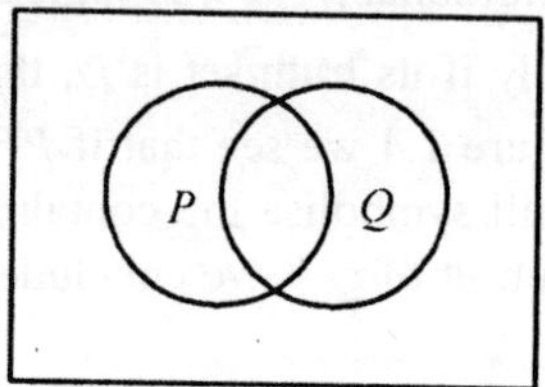

Figure: 1.11

The truth set of $p \vee (q \wedge r)$ is the entire shaded area of Figure 1.12a, and the truth set of $(p \vee q) \wedge (p \vee r)$ is the area in Figure 1.12b shaded in both colour and gray. Since these two sets are equal, we see that the two statements are equivalent.

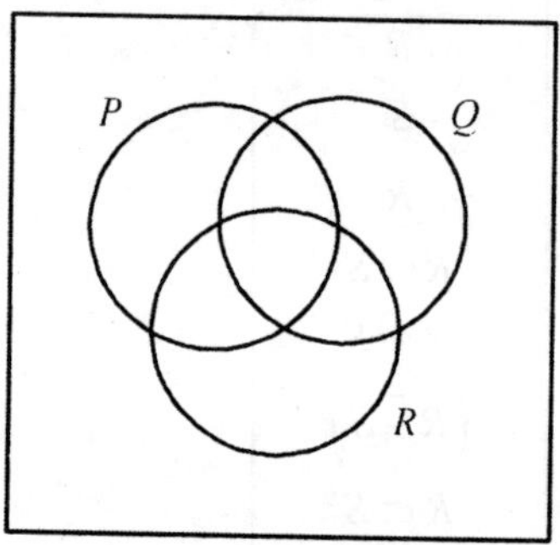

Figure: 1.12a

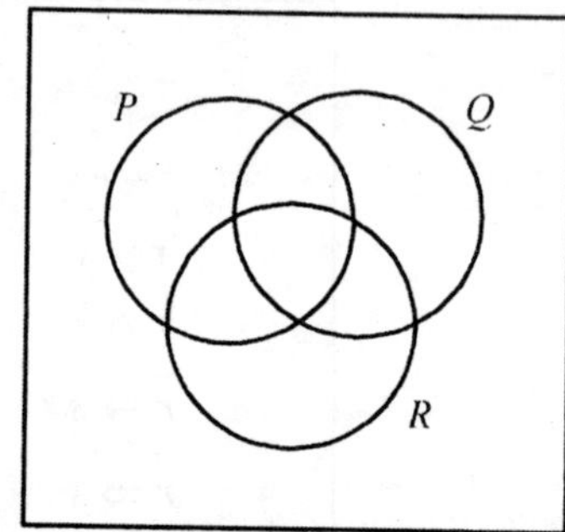

Figure: 1.12b

Example 3: Show by means of a Venn diagram that q implies $p \rightarrow q$. The truth set of $p \rightarrow q$ is the shaded area in Figure 1.9. Since this shaded area includes the set Q, we see that q implies $p \rightarrow q$.

Exercises

1. Use Venn diagrams to test the following statements for equivalences.

   (a) $\sim (p \vee q)$.

   (b) $\sim p \vee \sim q$.

(c) $\sim(p \wedge q)$.

(d) $\sim p \wedge \sim q$.

(e) $q \rightarrow p$.

(f) $\sim(\sim p \rightarrow q)$.

[*Ans.* (a), (d), and (f) are equivalent; (b) $\Leftrightarrow$ (c)]

2. Use Venn diagrams to tell which of the following statements are logically true and which are logically false.

   (a) $p \wedge \sim p$. [*Ans.* Logically true.]

   (b) $(p \wedge q) \vee (\sim p \vee \sim q)$.

   (c) $(p \wedge q) \vee (p \wedge \sim q)$.

   (d) $\sim p \vee (q \rightarrow p)$.

   (e) $p \rightarrow (q \rightarrow p)$.

   (f) $\sim(p \rightarrow q) \wedge q$.

3. Derive a test for inconsistency of *p* and *q*, using Venn diagrams.
4. Three or more statements are said to be inconsistent if they cannot all be true. What does this say about their truth sets?
5. Use Venn diagrams for the following statements to test whether one implies the other.

   (a) $p \wedge q$; $p \wedge \sim q$. (b) $\sim(q \rightarrow p)$; $p \rightarrow q$.

   (c) $p \wedge q$; $\sim p \vee q$. (d) $\sim p \wedge q$; q.

   (e) $p \vee q$; $p \rightarrow (\sim p \rightarrow q)$. (f) $(p \rightarrow q) \wedge \sim q$; $q \rightarrow p$.

6. Find statements having each of the following as truth sets.

   (a) $(P \cap Q) - R$.

   (b) $(R - Q) \cup (Q - R)$.

   (c) $P - (\widetilde{Q \cup R})$.

   (d) $(\widetilde{P \cap Q}) \cup (P \cup R)$.

7. Use truth tables to find whether the following sets are all different.

   (a) $(p \cap Q \cap \tilde{R}) \cup (P \cap \tilde{Q} \cap R) \cup (\tilde{P} \cap Q \cap R)$.

   (b) $[P - (Q \cup R)] \cup (R \cap Q)$.

(c) $Q \cap \tilde{R}$.

(d) $\left(P \cap Q \cap \tilde{R}\right) \cup \left(\tilde{P} \cap Q \cap \tilde{R}\right)$.

(e) $\left[(P \cap Q) \cup (P \cap R) \cup (r \cap Q)\right] - (p \cap Q \cap R)$.

(f) $\left[\left(P \cap \tilde{Q} \cap \tilde{R}\right) \cup \left[\overline{(Q \cup R) - (Q \cap R)}\right] - \left(\tilde{P} \cap \tilde{Q} \cap \tilde{R}\right)\right]$

8. Use truth tables to find whether each of the following sets is empty.

   (a) $(P - Q) \cap (Q - P)$. [*Ans.* Empty]

   (b) $(\tilde{P} \cup Q) \cap (\tilde{Q} \cup R) \cap (\tilde{P} \cup R)$.

   (c) $(\widetilde{P \cap R}) \cap (\tilde{P} \cap \tilde{Q})$. [*Ans.* Not empty]

   (d) $(\widetilde{P \cup R}) \cap \tilde{Q}$.

   (e) $(P \cap Q) - P$.

   (f) $(P \cap (Q - R)) - ((P \cap Q) - R)$.

9. Show, both by the use of truth tables and by the use of Venn diagrams, that $p \vee (q \wedge r)$ is equivalent to $(p \vee q) \wedge (p \vee r)$.

10. Use truth tables for the following pairs of sets to test whether one is a subset of the other.

    (a) $P \cap Q$; $\left[R - (\widetilde{P \cup Q})\right]$.

    (b) $(\tilde{P} \cap Q) \cap (\tilde{Q} \cup R)$; $\tilde{P} \cup R$.

    (c) $P \cap (Q \cup R)$; $P \cap Q$.

    (d) $P \cap \tilde{Q}$; $\tilde{P} \cap Q$.

    (e) Q; $\left(\tilde{P} \cup Q\right) \cap P$.

    (f) $P - (Q - R)$; $(P - Q) - R$.

11. The *symmetric difference* of P and Q is defined to be $(P - Q) \cup (Q - P)$. What connective corresponds to this set operation?

Permutations

The first step in the analysis of a scientific problem is the determination of the set of logical possibilities. Next, it is often necessary to determine how many different possible outcomes there are. We shall find this particularly important in probability theory. Hence, it is desirable to develop general techniques for solving counting problems. In this section and the next we shall discuss the two most important cases in which it is possible to achieve formulas that solve the problem. When a formula cannot be derived, one must resort to certain other general counting techniques, tricks, or, in the last resort, complete enumeration of the possibilities.

As a first problem, let us consider the number of ways in which a set of n different objects can be arranged. A *listing* of n different objects *in a certain order* is called a *permutation* of the n objects. We consider first the case of three objects, a, b, and c. We can exhibit all possible permutations of these three objects as paths of a tree, as shown in Figure 1.13.

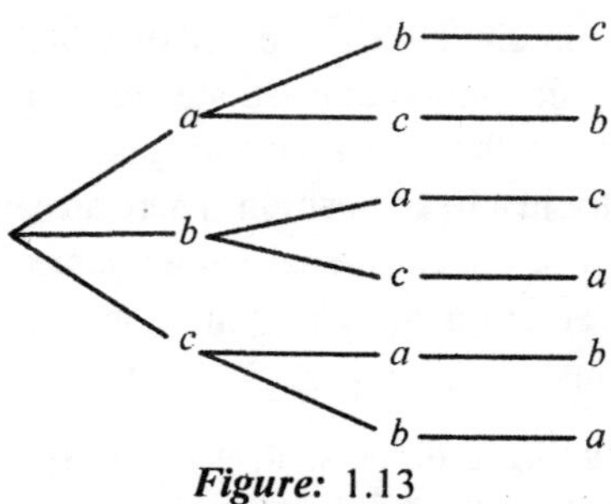

Figure: 1.13

Each path exhibits a possible permutation, and there are six such paths. We know there are six paths from the following argument: we have 3 choices for the first object; after this first choice, we can choose the second object in 2 ways; then the last object must be listed; thus the total number of listings is $3 \cdot 2 \cdot 1 = 6$. We could also list these permutations as follows:

abc, bca,

acb, cab,

bac, cba.

If we were to construct a similar tree for n objects, we would find that the number of paths could be found by multiplying together the numbers n, $n - 1$, $n - 2$, continuing down to the number 1. The number obtained in this way occurs so often that we give it a symbol, namely $n!$, which is read "n factorial." Thus, for example, $3! = 3 \cdot 2 \cdot 1 = 6$, $4!$

$= 4 \cdot 3 \cdot 2 \cdot 1 = 24$, and so on. For reasons that will be clear later, we define $0! = 1$. Thus, we can say *there are n! different permutations of n distinct objects.*

Example 1: Seven different machining operations are to be performed on a part, but they may be performed in any sequence. We may then consider $7! = 5040$ different orders in which the operations may be performed.

Example 2: Ten workers are to be assigned to 10 different jobs. In how many ways can the assignments be made? The first worker may be assigned in 10 possible ways, the second in any of the 9 remaining ways, the third in 8, and so forth: there are $10! = 3,628,800$ possible ways of assigning the workers to the jobs.

Example 3: A company has n directors. In how many ways can they be seated around a circular table at a board meeting, if two arrangements are considered different only if at least one person has a different person sitting on his right in the two arrangements? To solve the problem, consider one director in a fixed position. There are $(n - 1)!$ ways in which the other people may be seated. We have now counted all the arrangements we wish to consider different. Thus, there are also $(n - 1)!$ possible seating arrangements.

For many counting problems, it is not possible to give a simple formula for the number of possible cases. In many of these the only way to find the number of cases is to draw a tree and count them. In some problems, the following general principle is useful.

A General Principle: If one thing can be done in exactly r different ways, for each of these a second thing can be done in exactly s different ways, for each of the first two, a third can be done in exactly t ways, and so on, then the sequence of things can be done in $r \cdot s \cdot t \ldots$ ways.

Example 4: Suppose we live in town X and want to go to town Z by passing through town Y. If there are three roads from X to Y, and two roads from Y to Z, in how many ways can we go from town X to town Z? By applying the general principle we see that there are $3 \cdot 2 = 6$ ways.

The validity of this general principle can be established by thinking of a tree representing all the ways in which the sequence of things can be done. There would be r branches from the starting position. From the ends of each of these r branches there would be 5 new branches, and from each of these t new branches, and so on. The number of paths through the tree would be given by the product, $r \cdot s \cdot t \ldots$.

Example 5: The number of permutations of n distinct objects is a special case of this principle. If we were to list all the possible permutations, there would be n possibilities for the first, for each of these $n - 1$ for the second, etc., until we came to the last object, and for which there is only one possibility. Thus, there are $n(n - 1) \ldots 1 = n!$ possibilities in all.

Example 6: An automobile manufacturer produces four different models; models A and B can come in any of four body styles—sedan, hardtop, convertible, and station wagon—while models C and D come only as sedans or hardtops. Each can come in one of nine colours. Thus, models A and B each have $4 \cdot 9 = 36$ distinguishable types, while C and D have $2 \cdot 9 = 18$ types, so that in all

$$2 \cdot 36 + 2 \cdot 18 = 108$$

different car types are produced by the manufacturer.

Example 7: Suppose there are n applicants for a certain job. Three interviewers are asked independently to rank the applicants according to their suitability. It is decided that an applicant will be hired if he is ranked first by at least two of the three interviewers. What fraction of the possible reports would lead to the acceptance of some candidate? We shall solve this problem by finding the fraction of the reports that do not lead to an acceptance and subtract this answer from 1. Frequently, an indirect attack of this kind is easier than the direct approach. The total number of reports possible is $(n!)^3$, since each interviewer can rank the men in $n!$ different ways. If a particular report does not lead to the acceptance of a candidate, it must be true that each interviewer has put a different man in first place. By our general principle, this can be done in $n(n - 1)(n - 2)$ different ways. For each possible first choice, there are $[(n - 1)!]^3$ ways in which the remaining men can be ranked by the interviewers. Thus, the number of reports that do not lead to acceptance is

$$n(n - 1)(n - 2)[(n - 1)!]^3.$$

Dividing this number by $(n!)^3$, we obtain

$$\frac{(n-1)(n-2)}{n^2}$$

as the fraction of reports that fail to accept a candidate. The fraction that leads to acceptance is found by subtracting this fraction from 1, which gives

$$\frac{3n-2}{n^2}.$$

For the case of three applicants, we see that $\frac{7}{9}$ of the possibilities lead to acceptance. Here, the procedure might be criticised on the grounds that even if the interviewers are completely ineffective and are essentially guessing, there is a good chance that a candidate will be accepted on the basis of the reports. For n equal to ten, the fraction of acceptances is only .28, so that it is possible to attach more significance to the interviewers' ratings, if they reach a decision.

Exercises

1. A salesman is going to call on five customers. In how many different sequences can he do this if he...
   (a) Calls on all five in one day?
   (b) Calls on three one day and two the next?

   [*Ans.* (a) 120; (b) 120]

2. A machine shop has three milling machines, five lathes, six drill presses, and three grinders. In how many ways can a part be routed that must first be ground, then milled, then turned on a lathe, and then drilled? In how many ways can it be routed if these four operations can be performed in any order?

3. A department store wants to classify each of its customers having a charge account by using a three-character code consisting of n letters followed by 3 — n digits. How large must n be if there are 5000 charge accounts? What if there are 10,000? 20,000?

4. Modify Example 7 so that, to be accepted, an applicant must be first in two of the interviewers' ratings and must be either first or second in the third interviewer's rating. What fraction of the possible reports lead to acceptance in the case of three applicants? In the case of n?

   [*Ans.* $\frac{4}{9}$; $4/n^2$]

5. A company has six officers and six directors; two of the directors are officers. List the possible memberships of a committee of four men who are either officers or directors in terms of the number of members who are (a) just officers, (b) just directors, and (c) both officers and directors.

6. In Exercise 5, how many ways are there of obtaining a committee of four consisting of:

   (a) Three who are just officers and one who is officer and director?

   (b) One who is just an officer, one who is just a director, and two who are officers and directors?

   (c) At least two who are only directors and at least one who is officer and director?

   (d) At least two officers and at least two directors (assuming a man who is both officer and director satisfies both quotas)?

   [*Ans.* 160]

7. Show the possible arrangement of machines A, B, C, and D in a circle. How many are there?

8. How many possible ways are there of seating six people A, B, C, D, E, and F at a circular table if

   (a) A must always have B on his right and C on his left?

   (b) A must always sit next to B?

   (c) A cannot sit next to B?

9. In seating n people around a circular table, suppose we distinguish between two arrangements only if at least one person has at least one different person sitting next to him in the two arrangements. That is, we do not regard two arrangements as different simply because the right-hand and left-hand neighbours of a person have interchanged places. Now, how many distinguishable arrangements are there?

10. A certain symphony orchestra always plays one of the 41 Mozart symphonies, followed by one of 25 different modern works, followed by one of the 9 Beethoven symphonies.

    (a) How many different programmes can it play?

    (b) How many different programmes can be given if the pieces can be played in any order?

(c) How many three-piece programmes are possible if more than one piece from the same category can be played?

11. Find the number of arrangements of the five symbols that can be distinguished. (The same letters with different subscripts indicate distinguishable objects.)

(a) A_1, A_2, B_1, B_2, B_3. [*Ans.* 120]

(b) A, A, B_1, B_2, B_3. [*Ans.* 60]

(c) A, A, B, B, B. [*Ans.* 10]

12. Show that the number of distinguishable arrangements possible for n objects, n_1 of type 1, n_2 of type 2, and so on for r different types is

$$\frac{n!}{n_1!n_2!\ldots n_r!}.$$

13. A student takes a five-question multiple-choice test, each question having answer a, b, c, or d. If he knows that the answers to the test consist of two a's and one each of b, c, and d and he answers accordingly, in how many different ways can he answer the test? In what fraction of these will he get four or more right answers? In what fraction will he get three or more right? [*Partial Ans.* 60]

14. How many signals can a ship show if it has eight flags and a signal consists of five flags hoisted vertically on a rope?

[*Ans.* 6720]

15. We must arrange four green, one red, and four blue books on a single shelf. All books are distinguishable.

(a) In how many ways can this be done if there are no restrictions?

(b) In how many ways if books of the same colour must be grouped together?

(c) In how many ways if, in addition to the restriction in (b) the red books must be to the left of the blue books?

(d) In how many ways if, in addition to the restrictions in (b) and (c), the red and blue books must not be next to each other?

[*Ans.* 576]

16.(a) How many five-digit numbers can be formed from the digits 1, 2, 3, 4, 5 using each digit only once?

(b) How many of these numbers are less than 33,000?

17. A housewife who has just returned from shopping realises that she has left her sunglasses at either the bank, the post office, the drugstore, or the grocery store, and so she must go back and search for them. Assume that when she returns to the building where she left them, she finds them and then goes directly home.
    (a) In how many different orders can all four places be searched?
    (b) Assume we now know that she found her glasses at the third place she returned to. How many different searches can she have made?
    (c) If we know only that her glasses were left at the bank, how many different searches can she have made?

Binomial Coefficients

The binomial coefficients $\binom{n}{j}$ will play an important role in our future work. We give here some of the more important properties of these numbers.

A convenient way to obtain these numbers is given by the famous *Pascal triangle,* shown in Figure 1.14. To obtain the triangle we first write the l's down the sides. Any of the other numbers in the triangle has the property that it is the sum of the two adjacent numbers in the row just above. Thus, the next row in the triangle is 1,6, 15, 20, 15,6, 1. To find the binomial coefficient $\binom{n}{j}$ we look in the row corresponding to the number n and see where the diagonal line corresponding to the value of j intersects this row. For example, $\binom{4}{2} = 6$ is in the row marked $n = 4$ and on the diagonal marked $j = 2$.

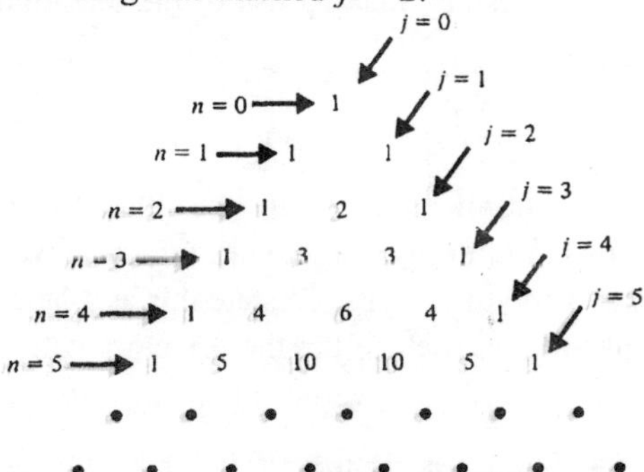

Figure: 1.14

The property of the binomial coefficients upon which the triangle is based is

$$\binom{n+1}{j}=\binom{n}{j-1}+\binom{n}{j}.$$

This fact can be verified directly, but the following argument is interesting in itself. The number $\binom{n+1}{j}$ is the number of subsets with y elements that can be formed from a set of $n + 1$ elements. Select one of the $n + 1$ elements, x. Of the $\binom{n+1}{j}$ subsets some contain x, and some do not. The latter are subsets of j elements formed from n objects, and hence there are $\binom{n}{j}$ such subsets. The former are constructed by adding x to a subset of $j - 1$ elements formed from n elements, and hence there are $\binom{n}{j-1}$ of them. Thus,

$$\binom{n+1}{j}=\binom{n}{j-1}+\binom{n}{j}.$$

If we look again at the Pascal triangle, we observe that the numbers in a given row increase for a while, and then decrease. In fact, they increase to a unique maximum when n is even or to two equal maxima when n is odd.

An important application of binomial coefficients is in the expansion of products of the form $(x + y)^3$, $(a - 2b)^{10}$, and so on. We shall derive a general formula for these by making use of the binomial coefficients.

Consider first the special case $(x + y)^3$. We write this as

$$(x + y)^3 = (x+ y)(x + y)(x + y).$$

To perform the multiplication, we choose either an x or y from each of the three factors and multiply our choices together; we do this for all possible choices and add the results. To state this as a labelling problem, note that we want to label each of the three factors with the two labels x and y. In how many ways can we do this using two x labels and one y? The preceding section gives the answer $\binom{3}{2}=3$. Hence, the coefficient of x^2y in the expansion of the binomial is 3. More generally, the coefficient of the term of the form x^jy^{3-j} will be $\binom{3}{j}$ for $j = 0, 1, 2, 3$. Thus, we can write the desired expansion as

$$(x+y)^3 = \binom{3}{3}x^3 + \binom{3}{2}x^2y + \binom{3}{1}xy^2 + \binom{3}{0}y^3$$
$$= x^3 + 3x^2y + 3xy^2 + y^3.$$

Binomial Theorem: The expansion of $(x + y)^n$ is given by

$$(x+y)^n = \binom{n}{n}x^n + \binom{n}{n-1}x^{n-1}y + \binom{n}{n-2}x^{n-2}y^2$$
$$+ \ldots + \binom{n}{1}xy^{n-1} + \binom{n}{0}y^n.$$

Example 1: Let us find the expansion for $(a - 2b)^3$. To fit this into the binomial theorem, we think of x as being a and y as being $-2b$. Then we have

$$(a - 2b)^3 = a^3 + 3a^2(-2b) + 3a(-2b)^2 + (-2b)^3$$
$$= a^3 - 6a^2b + 12ab^2 - 8b^3.$$

Exercises

1. Extend the Pascal triangle to $n = 16$. Save the result for later use.
2. (a) Show that a set with n elements has 2^n subsets. *[Hint:* Assume you have two different kinds of labels: "in the subset" and "not in the subset." In how many different ways can we label the n elements of the set?]

   (b) Prove that

$$\binom{n}{0} + \binom{n}{1} + \binom{n}{2} + \ldots + \binom{n}{n} = 2^n,$$

   using the fact that a set with n elements has 2^n subsets.
3. Using the fact that

$$\binom{n}{j+1} = \frac{n-j}{j+1}\binom{n}{j},$$

   compute $\binom{27}{s}$ for $s = 1, 2, 3, 4, 5$ starting with the fact that $\binom{27}{0} = 1$.
4. For $n \le m$ prove that

$$\binom{m}{0}\binom{n}{0}+\binom{m}{1}\binom{n}{1}+\binom{m}{2}\binom{n}{2}+\cdots+\binom{m}{n}\binom{n}{n}=\binom{m+n}{n}$$

by carrying out the following two steps:

(a) Show that the left-hand side counts the number of ways of choosing equal numbers of men and women from sets of m men and n women.

(b) Show that the right-hand side also counts the same number by showing that we can select equal numbers of men and women by selecting any subset of n persons from the whole set, and then combining the men selected with the women not selected.

5. Prove that

$$\binom{n+1}{j}=\binom{n}{j-1}+\binom{n}{j},$$

using only the fact that

$$\binom{n}{j}=\frac{n!}{j!(n-j)!}.$$

6. Expand by the binomial theorem:

(a) $(x+1)^4$. [*Ans.* $x^4 + 4x^3 + 6x^2 + 4x + 1$]

(b) $(2x+y)^3$.

(c) $(x-2)^5$.

(d) $(2a-x)^4$.

(e) $(3x+4y)^3$.

(f) $(100 - 2)^4$.

7. Using the binomial theorem, prove that

(a) $\binom{n}{0}+\binom{n}{1}+\binom{n}{2}+\cdots+\binom{n}{n}=2^n$.

(b) $\binom{n}{0}-\binom{n}{1}+\binom{n}{2}-\binom{n}{3}+\cdots\pm\binom{n}{n}$ 0 for $n > 0$.

Vectors and Matrices

Column and Row Vectors

A *column vector* is an ordered collection of numbers written in a column. Examples of such vectors are:

$$\begin{pmatrix} 1 \\ -2 \end{pmatrix}, \begin{pmatrix} .6 \\ .4 \end{pmatrix}, \begin{pmatrix} 0 \\ 0 \\ 0 \end{pmatrix}, \begin{pmatrix} 3 \\ -4 \\ 0 \end{pmatrix}, \begin{pmatrix} 1 \\ -1 \\ 2 \\ 4 \end{pmatrix}.$$

The individual numbers in these vectors are called *components*, and the number of components a vector has is one of its distinguishing characteristics. Thus, the first two vectors above have two components; the next two have three components; and the last has four components. When talking more generally about *n*-component column vectors we shall write

$$u = \begin{pmatrix} u_1 \\ u_2 \\ \cdot \\ \cdot \\ \cdot \\ u_n \end{pmatrix}.$$

Analogously, a *row vector* is an ordered collection of numbers written in a row. Examples of row vectors are

$$(1, 0), (-2, 1), (2, -3, 4, 0), (-1, 2, -3, 4, -5).$$

Each number appearing in the vector is again called a *component of the vector*, and the number of components a row vector has is again one of its important characteristics. Thus, the first two examples are two-component, the third a four-component, and the fourth a five-component vector. The vector $v = (v_1, v_2, \ldots, v_n)$ n-component row vector.

Two row vectors or two column vectors, are said to be equal if and only if corresponding components of the vector are equal. Thus, for the vectors

$$u = (1, 2),\quad v = \begin{pmatrix} 1 \\ 2 \end{pmatrix},\quad w = (1, 2),\ x = (2, 1)$$

we see that $u = w$ but $u \neq v$, and $u \neq x$.

If u and v are three-component column vectors, we shall define their sum $u + v$ by component wise addition as follows:

$$u + v = \begin{pmatrix} u_1 \\ u_2 \\ u_3 \end{pmatrix} + \begin{pmatrix} v_1 \\ v_2 \\ v_3 \end{pmatrix} = \begin{pmatrix} u_1 + v_1 \\ u_2 + v_2 \\ u_3 + v_3 \end{pmatrix}$$

Similarly, if u and v are three-component row vectors, their sum is defined to be

$$\begin{aligned} u + v &= (u_1, u_2, u_3) + (v_1, v_2, v_3) \\ &= (u_1 + v_1, u_2 + v_2, u_3 + v_3) \end{aligned}$$

Note that the sum of two three-component vectors yields another three-component vector. For example,

$$\begin{pmatrix} 1 \\ -1 \\ 2 \end{pmatrix} + \begin{pmatrix} 2 \\ 3 \\ -1 \end{pmatrix} = \begin{pmatrix} 3 \\ 2 \\ 1 \end{pmatrix}$$

and

$$(4, -7, 12) + (3, 14, -14) = (7, 7, -2).$$

The sum of two n-component vectors (either row or column) is defined by component wise addition in an analogous manner, and yields another n-component vector. Observe that we do not define the addition of vectors unless they are both row or both column vectors having the same number of components.

Because the order in which two numbers are added does not affect the answer, it is also true hat the order in which vectors are added does not matter; that is,

$$u + v = v + u,$$

where u and v are both row or both column vectors. This is the so-called *commutative law of addition.* A numerical example is

$$\begin{pmatrix} 1 \\ -1 \\ 2 \end{pmatrix} + \begin{pmatrix} 2 \\ 3 \\ -1 \end{pmatrix} = \begin{pmatrix} 3 \\ 2 \\ 1 \end{pmatrix} = \begin{pmatrix} 2 \\ 3 \\ -1 \end{pmatrix} + \begin{pmatrix} 1 \\ -1 \\ 2 \end{pmatrix}$$

Once we have the definition of the addition of two vectors, we can easily se how to add three or more vectors by grouping them in pairs as in the addition of numbers. For example,

$$\begin{pmatrix} 1 \\ 0 \\ 0 \end{pmatrix} + \begin{pmatrix} 0 \\ 2 \\ 0 \end{pmatrix} + \begin{pmatrix} 0 \\ 0 \\ 3 \end{pmatrix} = \begin{pmatrix} 1 \\ 0 \\ 0 \end{pmatrix} + \begin{pmatrix} 0 \\ 2 \\ 3 \end{pmatrix} = \begin{pmatrix} 1 \\ 2 \\ 3 \end{pmatrix} = \begin{pmatrix} 1 \\ 2 \\ 0 \end{pmatrix} + \begin{pmatrix} 0 \\ 0 \\ 3 \end{pmatrix} = \begin{pmatrix} 1 \\ 2 \\ 3 \end{pmatrix}$$

and

$$(1, 0, 0) + (0, 2, 0) + (0, 0, 3) = (1, 2, 0) + (0, 0, 3) = (1, 2, 3)$$
$$= (1, 0, 0) + (0, 2, 3) = (1, 2, 3).$$

In general, the sum of any number of vectors (row or column), each having the same number of components, is the vector whose first component is the sum of the first components of the vectors, whose second component is the sum of the second components, and so on.

The multiplication of a number a times a vector v is defined by component wise multiplication of a times the components of v. For the three-component case we have

$$au = a\begin{pmatrix} u_1 \\ u_2 \\ u_3 \end{pmatrix} = \begin{pmatrix} au_1 \\ au_2 \\ au_3 \end{pmatrix}$$

for column vectors and

$$av = a(v_1, v_2, v_3) = (av_1, av_2, av_3)$$

for row vectors. If u is an n-component vector (row or column), then au is defined similarly by component wise multiplication. This operation is sometimes called *scalar multiplication* of a vector, where scalar is another name for a number.

If u is any vector, we define its negative $-u$ to be the vector $-u = (-1)u$. Thus, in the three-component case for row vectors we have

$$-u = (-1)(u_1, u_2, u_3,) = (-u_1, -u_2, -u_3).$$

Once we have the negative of a vector it is easy to see how to subtract vectors: we simple add "algebraically." For the three-component column-vector case we have

$$u - v = \begin{pmatrix} u_1 \\ u_2 \\ u_3 \end{pmatrix} - \begin{pmatrix} v_1 \\ v_2 \\ v_3 \end{pmatrix} = \begin{pmatrix} u_1 - v_1 \\ u_2 - v_2 \\ u_3 - v_3 \end{pmatrix}.$$

Specific examples of subtraction of vectors occur in the exercises at the end of this section.

An important vector is the zero vector, all of whose components are zero. For example, three-component zero vectors are

$$0 = \begin{pmatrix} 0 \\ 0 \\ 0 \end{pmatrix} \quad \text{and} \quad 0 = (0, 0, 0).$$

When there is no danger of confusion we shall use the symbol 0, as above, to denote the zero (row or column) vector. The meaning will be clear from the context. The zero vector has the important property that, if u is any vector, then $u + 0 = u$. A proof for the three-component column-vector case is as follows:

$$u + 0 = \begin{pmatrix} u_1 \\ u_2 \\ u_3 \end{pmatrix} + \begin{pmatrix} 0 \\ 0 \\ 0 \end{pmatrix} = \begin{pmatrix} u_1 + 0 \\ u_2 - 0 \\ u_3 - 0 \end{pmatrix} = \begin{pmatrix} u_1 \\ u_2 \\ u_3 \end{pmatrix} = u.$$

One of the chief advantages of the vector notation is that we can denote a whole collection of numbers by a single letter such as u, v,..., and treat such a collection as if it were a single quantity. By using the vector notation we can state very complicated relationships in a simple manner. The student will see many examples of this in the remainder of the present chapter.

Exercises

1. Compute the quantities below for the vectors

$$u = \begin{pmatrix} 2 \\ 1 \\ 8 \\ 6 \end{pmatrix}, \; v = \begin{pmatrix} 3 \\ -5 \\ 2 \\ 0 \end{pmatrix}, \; w = \begin{pmatrix} 6 \\ 6 \\ -6 \\ -6 \end{pmatrix}.$$

   (a) $u + w$. [*Ans.* $\begin{pmatrix} 8 \\ 7 \\ 2 \\ 0 \end{pmatrix}$]

   (b) $5w$.

   (c) $v - u$.

   (d) $3u + 7v - 2w$. [*Ans.* $\begin{pmatrix} 15 \\ -44 \\ 50 \\ 30 \end{pmatrix}$]

   (e) $\frac{1}{2}w + \frac{3}{4}u$.

   (f) $u - w - v$.

   (g) $-2u + 3v - 100w$.

2. Compute (a) through (g) of Exercise 1 if the vectors u, v and w are: $u = (7, 0, -3)$, $v = (2, 1, -5)$, $w = (1, -1, 0)$.

3. (a) Show that the zero vector is not is not changed when multiplied by any number.

   (b) If u is any vector, show that $0 + u = u$.

4. If $2u - v = 0$, what is the relationship between the components of u and those of v? [*Ans.* $v_i = 2u_i$]

5. When possible, compute the following sums; when not possible, give reasons.

   (a) $(2, 3) + 2\begin{pmatrix} 1 \\ 5 \end{pmatrix} = ?$

(b) $0(5, 1, 7) + 3(6, 2, 6) = ?$

(c) $\begin{pmatrix} 3 \\ 1 \\ 2 \end{pmatrix} + 5 + \begin{pmatrix} 6 \\ 2 \\ 6 \end{pmatrix} = ?$ [*Ans.* Not possible]

(d) $\begin{pmatrix} 21 \\ 22 \\ 23 \\ 24 \end{pmatrix} + 32 \begin{pmatrix} 1 \\ 2 \\ 0 \\ 1 \end{pmatrix} = ?$ $\left[Ans. \begin{pmatrix} 53 \\ 86 \\ 23 \\ 56 \end{pmatrix}\right]$

6. If $\begin{pmatrix} 6 \\ 6 \\ 0 \end{pmatrix} + \begin{pmatrix} u_1 \\ u_2 \\ u_3 \end{pmatrix} = \begin{pmatrix} 5 \\ -5 \\ 5 \end{pmatrix}$, find u_1, u_2, and u_3. [*Ans.* –1, –11, 5]

7. If $8 \begin{pmatrix} v_1 \\ v_2 \\ v_3 \end{pmatrix} = \begin{pmatrix} 1 \\ -16 \\ 0 \end{pmatrix}$, find the components of v.

8. Find three vectors u, v and w such that $w = 3u$, $v = 2u$, and $2u + 3v + 4w = \begin{pmatrix} 20 \\ 10 \\ -25 \end{pmatrix}$.

9. If $\begin{pmatrix} 0 \\ 0 \\ 0 \end{pmatrix} + \begin{pmatrix} u_1 \\ u_2 \\ u_3 \end{pmatrix} = \begin{pmatrix} 0 \\ 0 \\ 0 \end{pmatrix}$, what can be said concerning the components u_1, u_2, u_3?

10. If $0 \begin{pmatrix} u_1 \\ u_2 \\ u_3 \end{pmatrix} = \begin{pmatrix} 0 \\ 0 \\ 0 \end{pmatrix}$, what can be said concerning the components u_1, u_2, u_3?

11. If $(u_1 + u_2 + u_3 + u_4) \begin{pmatrix} u_1 \\ u_2 \\ u_3 \\ u_4 \end{pmatrix} = \begin{pmatrix} 1 \\ 3 \\ 5 \\ 7 \end{pmatrix}$, what are u_1, u_2, u_3, and u_4?

12. (a) Show that the vector equation

$$x\begin{pmatrix}5\\7\end{pmatrix}+y\begin{pmatrix}3\\-10\end{pmatrix}=\begin{pmatrix}-16\\77\end{pmatrix}$$

represents two simultaneous linear equations for the two variables x and y.

(b) Solve the equations for x and y from (a) and substitute into the vector equation above to check your work.

13. Write the following simultaneous linear equations in vector form:

$ax + by = e$

$cx + dy = f.$

[*Hint:* Follow the form given in Exercise 12.]

14. Suppose that we associate with each person a three-component row vector having the following entries: age, height and weight. Would it make sense to add together the vectors associated with two different persons? Would if make sense to multiply one of these vectors by a constant?

15. Suppose that we associate with each person leaving a supermarket a row vector whose components give quantities of each available item that he has purchased. Answer the same questions as those in Exercise 14.

16. Let us associate with each supermarket a vector whose entries give the prices of each item in the store. Would it make sense to add together the vectors associated with two different supermarkets? Would it make sense to multiply one of these vectors by a constant? Discuss the differences in the situations given in Exercises 14, 15, and 16.

17. Consider the vectors

$$x=\begin{pmatrix}x_1\\x_2\end{pmatrix},\quad y=\begin{pmatrix}y_1\\y_2\end{pmatrix}.$$

Show that the vector

$\frac{1}{2}(x+y)$

has components that are the averages of the components of x and y. Generalise this result to the case of n vectors.

18. Would the concept of averages, as discussed in Exercise 17, be applicable to the vectors mentioned in Exercises 14, 15, and 16? How would the averages be interpreted?

19. In a certain school, students take four courses each semester. At the end of the semester the registrar recorded the grades of each student as a row vector. He then gives the student 4 points for each A, 3 points for each B, 2 points for each C, 1 point for each D, and 0 for each F. The sum of these numbers, divided by 4, is the student's grade point average.

 (a) If a student has a 4.0 average, what are the logical possibilities for his grade vector? [*Hint:* Each grade vector will have five components.]

 (b) What are the possibilities if he has a 3.0 average?

 (c) What are the possibilities if he has a 2.0 average?

20. Let $x = \begin{pmatrix} x_1 \\ x_2 \end{pmatrix}$. Define $x \geq 0$ to be the conjunction of the statements $x_1 \geq 0$ and $x_2 \geq 0$. Define $x \geq 0$ analogously. If $(x_1 + x_2)x \geq 0$, what must be true of *x*?

21. Using the definition in Exercise 20, define $x \geq y$ to mean $x - y \geq 0$, where *x* and *y* are vectors of the same shape. Consider the following four vectors:

$$x = \begin{pmatrix} 3 \\ 5 \\ -1 \end{pmatrix}, \quad y = \begin{pmatrix} 6 \\ 5 \\ 6 \end{pmatrix}, \quad u = \begin{pmatrix} 0 \\ 0 \\ 2 \end{pmatrix}, \quad v = \begin{pmatrix} 4 \\ 2 \\ 0 \end{pmatrix}.$$

 (a) Show that $x \geq u$.

 (b) Show that $v \geq u$.

 (c) Is there any relationship between *x* and *v*?

 (d) Show that $y \geq x$, $y \geq u$, and $y \geq v$.

22. (a) If $x^{(1)}, x^{(2)}, \ldots, x^{(n)}$ is a set of *n* vectors, show how to find a vector *u* such that $u \geq x^{(i)}$ for all *i*. Also show how to find a vector *v* such that $v \leq x^{(i)}$ for all *i*.

 (b) Apply the results of part (a) to the vectors in Exercise 21.

(c) Let $u = \begin{pmatrix} 630 \\ 520 \\ 310 \end{pmatrix}$, $v = \begin{pmatrix} 960 \\ 200 \\ 400 \end{pmatrix}$, and $w = \begin{pmatrix} 600 \\ 750 \\ 490 \end{pmatrix}$. If $x = \begin{pmatrix} -4 \\ -5 \\ -2 \end{pmatrix}$, find the largest number n such that $nx \geq u$, $nx \geq v$, and $nx \geq w$.

[*Ans.* –235]

Product of Vectors

The reader may wonder why it is necessary to introduce both column and row vectors when their properties are so similar. This question can be answered in several different ways. First, in many applications two kinds of quantities are studied simultaneously, and it is convenient to represent one of them as a row vector and the other as a column vector. Second, there is a way of combining row and column vectors that is very useful for certain types of calculations. To bring out these points let us look at the following simple economic example.

Example 1: Suppose a man named Smith goes into a grocery store to buy a dozen each of peaches and oranges, a half-dozen each of apples and pears, and three lemons. Let us represent his purchases by means of the following row vector:

$$x = [6\text{ (apples)}, 12\text{ (peaches)}, 3\text{ (lemons)}, 12\text{ (oranges)}, 6\text{ (pears)}]$$

$$= (6,\ 12,\ 3,\ 12,\ 6).$$

Suppose that apples are 4 cents each, peaches 6 cents, lemons 9 cents, oranges 5 cents, and pears 7 cents. We can then represent the prices of these items as a column vector:

$$y = \begin{pmatrix} 4 \\ 6 \\ 9 \\ 5 \\ 7 \end{pmatrix} \begin{matrix} \text{cents per apple} \\ \text{cents per peach} \\ \text{cents per lemon} \\ \text{cents per orange} \\ \text{cents per pear.} \end{matrix}$$

The obvious question is: What is the total amount that Smith must pay for his purchases? We would like to multiply the quantity vector x by the price vector y, and we would like the result to be Smith's bill. We see that our multiplication should have the following form:

$$x \cdot y = (6,12,3,12,6)\begin{pmatrix}4\\6\\9\\5\\7\end{pmatrix}$$

$$= 6 \cdot 4 + 12 \cdot 6 + 3 \cdot 9 + 12 \cdot 5 + 6 \cdot 7$$

$$= 24 + 72 + 27 + 60 + 42$$

$$= 225 \text{ cents or } \$2.25.$$

This is, of course, the computation that the cashier performs in figuring Smith's bill.

We shall adopt in general the above definition of multiplication of row times column vectors.

Definition: Let u be a row vector and v a column vector each having the same number n of components; than we shall define the produce $u \,.\, v$ to be

$$u \cdot v = u_1 v_1 + u_2 v_2 + \ldots + u_n v_n.$$

Notice that we always write the row vector first and the column vector second, and this is the only kind of vector multiplication that we consider. Some examples of vector multiplication are

$$(2,1,-1) \cdot \begin{pmatrix}3\\-1\\4\end{pmatrix} = 2 \cdot 3 + 1 \cdot (-1) + (-1) \cdot 4 = 1,$$

$$(1, 0) \cdot \begin{pmatrix}0\\1\end{pmatrix} = 1 \cdot 0 + 0 \cdot 1 = 0 + 0 = 0.$$

Note that the result of vector multiplication is always a *number*.

Example 2: Consider an oversimplified economy that has three industries, which we coal, electricity, and steel and three consumers 1, 2, and 3. Suppose that each consumer uses some of the output of each industry and also that each industry uses some of the output of each other industry. We assume that the amounts used are positive or zero, since using a negative quantity has no immediate interpretation. We can represent the needs of each consumer and industry by a three-component demand (row) vector, the first component measuring the amount of coal needed by the consumer or industry; the second component the amount of electricity needed; and the third component the amount of steel needed, in some convenient units. For example, the demand vectors of the three consumers might be

$$d_1 = (3,\ 2,\ 5),\ d_2 = (0,\ 17,\ 1),\ d_3 = (4,\ 6,12)$$

and the demand vectors of each of the industries might be

$$d_C = (0,\ 1,\ 4),\ d_E = (20,\ 0,\ 8),\ d_S = (30,\ 5,\ 0),$$

where the subscript C stands for coal, the subscript E for electricity, and the subscript S for steel. Then the total demand for these goods by the consumers is given by the sum

$$d_1 + d_2 + d_3 = (3,\ 2,\ 5) + (0,\ 17,\ 1) + (4,\ 6,\ 12) = (7,\ 25,\ 18).$$

Also, the total industrial demand for these goods is given by the sum

$$d_C + D_E + d_S = (0,1,4) + (20,0,8) + (30,5,0) = (50,6,12)$$

Therefore the total overall demand is given by the sum

$$(7,\ 25,\ 18) + (50,\ 6,\ 12) = (57,\ 31,\ 30).$$

Suppose now that the price of coal is \$1 per unit, the price electricity is \$2 per unit, and the price of steel is \$4 per unit. Then these prices can be represented by the column vector

$$p = \begin{pmatrix} 1 \\ 2 \\ 4 \end{pmatrix}.$$

Consider the steel industry: it sells a total of 30 units of steel at \$4 per unit, so that its total income is \$120. Its bill for the various goods is given by the vector product

$$d_S \cdot p = (30,5,0) \cdot \begin{pmatrix} 1 \\ 2 \\ 4 \end{pmatrix} = 30 + 10 + 0 = \$40.$$

Hence, the profit of the steel industry is \$120 \$40 = \$80. (In the exercises the profits of the other industries will be found.)

This model of an economy is unrealistic in two senses. First, we have not chosen realistic numbers for the various quantities involved. Second, and more important, we have neglected the fact that the more an industry produces the more inputs it requires.

Example 3: Consider the rectangular coordinate system in the plane shown in Figure 2.1. A two-component row vector $x = (a, b)$ can be regarded as a point in

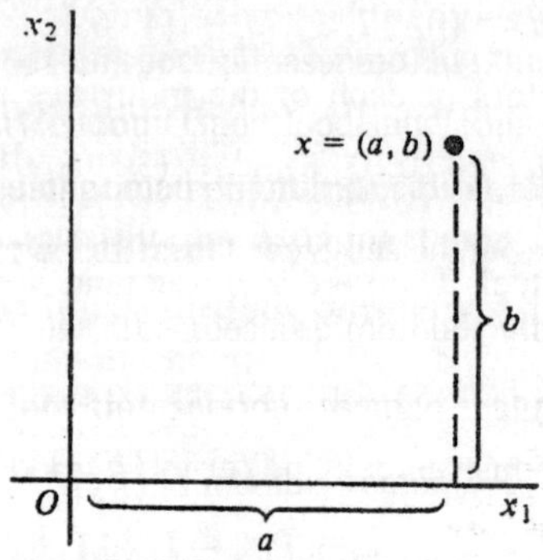

Figure: 2.1

the plane located by means of the coordinate axes as shown. The point x can be found by starting at the origin of coordinates O and moving a distance a along the x_1 axis; then moving a distance b along a line parallel to the x_2 axis. If we have two such points, say $x = (a, b)$ and $y = (c, d)$, then the points $x + y$, $-x$, $-y$, $x - y$, $y - x$, $-x - y$ have the geometric significance shown in Figure 2.2.

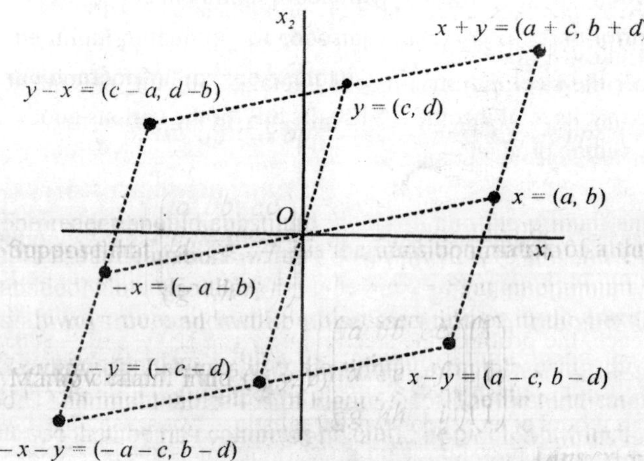

Figure: 2.2

The idea of multiplying a row vector by a number can also be given a geometric meaning. In Figure 2.3 we have plotted the point corresponding to the vector $x = (1, 2)$, and $2x$, $\frac{1}{2}x$, $-x$, and $-2x$. Observe that all these points lie on a line through the origin of coordinates. Another vector quantity that has geometrical significance is the vector $z = ax + (1 - a)y$,

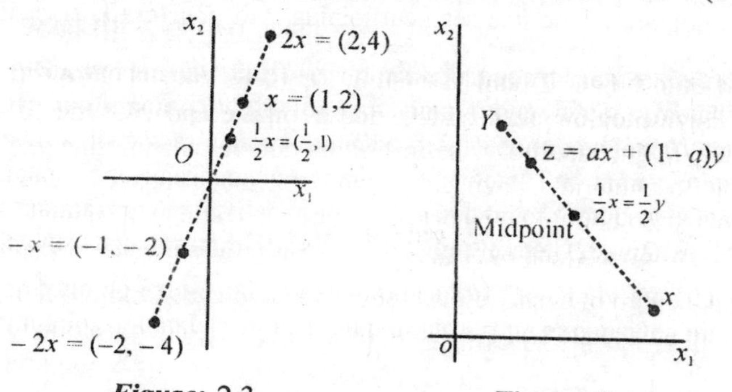

Figure: 2.3 ***Figure:*** 2.4

where a is any number between 0 and 1. Observe in Figure 2.4 that the points z all lie on the line segment between the points x and y. If $a = \frac{1}{2}$, the corresponding point on the line segment is the midpoint of the segment. Thus, if $x = (a, b)$ and $y = (c, d)$, then the point

$$\frac{1}{2}x + \frac{1}{2}y = \frac{1}{2}(a, b) + \frac{1}{2}(c, d)$$

$$= \left(\frac{a+c}{2}, \frac{b+d}{2}\right)$$

is the midpoint of the line segment between x and y.

Exercises

1. Let $u = (2, -1, 3)$, $V = (5, 0, 2)$, $x = \begin{pmatrix} -7 \\ 1 \\ 2 \end{pmatrix}$, and $y = \begin{pmatrix} 0 \\ 8 \\ 3 \end{pmatrix}$

Compute the following:

(a) $(u + v) \cdot (x + y)$.

(b) $(3u \cdot x) \cdot v) \cdot y$.

(c) $u \cdot x - 4v \cdot y \cdot$

(d) $u \cdot x + 3u \cdot y - u \cdot y$. [Ans. —12]

(e) $(2(v + u) \cdot y) - 5uy$.

(f) $4u \cdot x + 6[v \cdot (3x - y)]$ [Ans. —630]

2. If $x = (5, -4, 2)$ and $y = (1,8,1)$ are points in space, what is the midpoint of the line segment joining x to y? [Ans.$(3, 2, \frac{1}{2})$]

3. Let $x = (-1, 3)$ and $y = (4, 1)$ be Row Vectors. Plot the points corresponding to x and y, and compute and plot the following vectors:

(a) $\frac{1}{2}x+\frac{1}{2}y$. (d) $2x-y$.

(b) $x+y$. (e) $y+4x$.

(c) $\frac{2}{3}x+\frac{1}{3}y$. (f) $x-y$.

4. Prove that vector multiplication satisfies the following two properties:

(i) $u \cdot (av) = a(u \cdot v)$,

(ii) $u \cdot (v + w) = u \cdot v + u \cdot w$,

where u is a three-component row vector, v and w are three-component column vectors, and a is a number.

5. If u is a three-component row vector, v is a three-component column vector having the same number of components, and a is a number prove that $a(u \cdot v) = u \cdot (av)$.

6. A certain football stadium has three gates. After one game, the ticket-taker at gate 1 reported admitting 275 adults, 300 students, and 15 children; the ticket-taker at gate 2 admitted 200 adults, 107 students, and 40 children; and 65 adults, 250 students, and 60 children were admitted at age 3.

(a) Write the numbers of people admitted through each gate as three-component row vector.

(b) Use vector addition to find how many people in each category attend the game.

(c) Suppose that adults pay $3.00 to attend the game, students pay $2.00, and children pay ¢50. Assuming that each person buys his ticket at the gate, calculate the value of the tickets sold at each gate.

(d) Compute in two different ways the total value of tickets sold.

7. Perform the following calculations for Example 2.

   (a) Compute the amount that each industry and each consumer has to pay for the good it receives.

   (b) Compute the profit made by each of the industries.

   (c) Find the total amount of money that is paid out by all the industries and consumers.

   (d) Find the proportion of the total amount of money found in (c) paid out by the industries. Find the proportion of the total money that is paid out the consumers.

8. A farmer intends to plant 40 acres of corn, 25 acres of wheat, and 30 acres of rye. Write a three-component row vector whose components give the number of acres of each grain he wants to plant. Suppose an acre of corn requires an hour to plant, an acre of wheat 45 minutes, and an acre of rye one-half hour. Write a column vector whose components give the number of minutes needed to plant an acre of each crop. Use vector multiplication to find the total time required for planting.

9. In Exercise 8, suppose the seed for an acre of corn cost \$40 for an acre of wheat \$15, and for an acre of rye \$10. Find the total amount the farmer must spend on seed. If an acre of corn brings the farmer a profit of \$19, an acre of wheat \$15, and an acre of rye \$4, find the total profit the farmer will make.

10. The production of a book involves several steps: first it must be set into type, then it must be printed, and finally it must be supplied with covers and bound. Suppose the typesetter charges \$6 an hour, paper costs $\frac{1}{4}$ ¢ per sheet, the printer charges 11¢ for each minute that his press runs, the cover costs 28¢, and the binder charges 15 cents to bind each book. Suppose now that a publisher wishes to print a book that requires 300 hours of work by the typesetter, 220 sheets of paper per book, and 5 minutes of press time per book.

    (a) Write a five-component row vector that gives the requirements for the first book. Write another row vector that gives the requirements for the second, third, ... copies of the book. Write a five-component column vector whose components give the prices of the various requirements for each book, in the same order as they are listed in the requirement vectors above.

(b) Using vector multiplication, find the cost of publishing one copy of book. [*Ans.* \$1801.53]

(c) Using vector addition and multiplication, find the cost printing a first-edition run of 5,000 copies.[*Ans.* \$9450]

(d) Assuming that the printing plates from the first edition are use again, find the cost of printing a second edition of 5,000 copies. [*Ans.* \$7,650.]

11. Let $x = \begin{pmatrix} x_1 \\ x_2 \end{pmatrix}$, and let a and b be the vectors $a = (-1, 4)$, $b = (2, -7)$. If $ax = 1$ and $bx = 2$, find x_1 and x_2

[*Ans.* $x_1 = 15, x_2 = 4$]

12. Let $x = \begin{pmatrix} x_1 \\ x_2 \end{pmatrix}$, and let a and b be the vectors $a = (3, 3)$, $b = (-1, 4)$. If $ax = x_1$ and $bx = x_2$, find x_1 and x_2.

13. Consider the vectors

$$a = (a_1, a_2),\ b = (b_1, b_2),\ x = \begin{pmatrix} x_1 \\ x_2 \end{pmatrix}$$

and two number c_1 and c_2. Show that the equations

$ax = c_1,\ bx = c_2$

represent two simultaneous equations in two unknowns.

14. Show that every set of two simultaneous equations in two unknowns can be written as in Exercise 13.

15. Consider an experiment in which there are two outcomes: we win \$10 with probability $\frac{1}{5}$ and \$ with probability $\frac{4}{5}$. Let $a = (10, -5)$ and $p = \begin{pmatrix} \frac{1}{5} \\ \frac{4}{5} \end{pmatrix}$. Show that the expected outcome of the experiment is ap.

16. A gambling game works as follows. Two dice are rolled. If no sixes turn up, we lose \$1; if one six turns up, we win \$1; if two sixes

turn we win \$10. Set up a row vector representing the various outcomes and a column vector representing the probability of those outcomes. Use vector multiplication to find the expected outcomes. Is the game fair?

17. If an experiment has outcomes $a_1, a_2, ... a_n$ occurring with probabilities $p_1, p_2, ..., p_n$, define the vectors

$$a = (a_1, ..., a_n) \text{ and } p = \begin{pmatrix} p_1 \\ p_2 \\ \vdots \\ p_n \end{pmatrix}$$

Show that the expected outcome is ap.

18 Consider the vectors $x = (1, 5)$, $y = (3, 1)$ and $f = \begin{pmatrix} 1 \\ 1 \end{pmatrix}$.

(a) Compute $\frac{1}{2}xf$ and $\frac{1}{2}yf$, and show that these numbers are the averages of the components of x and y respectively.

[*Partial Ans.* 3, 2.]

(b) Compute $\frac{1}{4}(x+y)f$, and given an interpretation for this number.

[*Partial Ans.*$2\frac{1}{2}$]

19. Let x and y be two n-component Row Vectors, and let f be an n-component column vector all whose entries are 1's.

(a) Compute $(1/n)\ xf$ and $(1/n)yf$ and interpret the result.

(b) Compute $(\frac{1}{2}n)\ (x + y)f$ and interpret the result. [*Hint*: Exercise 18 is a special case.]

20. How would the results of Exercise 19 change if we used three vectors: x, y, and z?

Matrices and their Combination with Vectors

A matrix is a rectangular array of numbers written in the from

$$A = \begin{pmatrix} a_{11} & a_{12} & \cdots & a_{1n} \\ a_{21} & a_{22} & \cdots & a_{2n} \\ . & . & \cdots & . \\ a_{m1} & a_{m2} & \cdots & a_{mn} \end{pmatrix}$$

Here, the letters stand for real number and m and n are integers. Observe that m is the number of rows and n is the number of columns of the matrix, For this reason we call it an $m \times n$ matrix. The following are examples of matrices:

$$(1, 2, 3), \begin{pmatrix} 1 \\ 2 \\ 3 \end{pmatrix}, \begin{pmatrix} \frac{1}{-2} & \frac{-1}{2} \end{pmatrix};$$

$$\begin{pmatrix} 1 & 0 & 0 & 0 \\ 0 & 1 & 0 & 0 \\ 0 & 0 & 1 & 0 \\ 0 & 0 & 0 & 1 \end{pmatrix}; \begin{pmatrix} 1 & 7 & -8 & 9 & 10 \\ 3 & -1 & 14 & 2 & -6 \\ 0 & 3 & -5 & 7 & 0 \end{pmatrix}.$$

The first example is a row vector which is a 1×3 matrix; the second is a column vector which is a 3×1 matrix; the third example is a 2×2 square matrix; the fourth is a 4×4 square matrix; and the last is a 3×5 matrix.

Two matrices having the same shape (i.e., having the same number of rows and columns) are said to be equal if and only if the corresponding entries are equal.

Recall that we found a matrix arose naturally in the consideration of a Markov chain process. To give another example of how matrices occur in practice and are used in connection with vectors, we consider the following example.

Example 1: Suppose that a building contractor has accepted orders for five ranch-style houses, seven Cape Cod houses, and twelve colonial-style houses. We can represent his orders by means of a row vector x = (5, 7, 12). The contractor is familiar, of course, with the kinds of "raw materials" that go into each type of house. Let us suppose that these raw materials are steel, wood, glass, paint, and labour. The numbers in the matrix below give the amounts of each raw material going into each type of house, expressed in convenient units. (The numbers are put in arbitrarily, and are not meant to be realistic) .

$$\begin{array}{r} \\ \text{Ranch:} \\ \text{Cape Cod:} \\ \text{Colonial:} \end{array} \begin{array}{c} \begin{array}{ccccc} \text{Steel} & \text{Wood} & \text{Glass} & \text{Paint} & \text{Labor} \end{array} \\ \begin{pmatrix} 5 & 20 & 16 & 7 & 17 \\ 7 & 18 & 12 & 9 & 21 \\ 6 & 25 & 8 & 5 & 13 \end{pmatrix} \end{array} = R.$$

Observe that each row of the matrix is a five-component row vector which give the amounts of each raw material needed for a given kind of house. Similarly, each column of the matrix is a three-component column vector which gives the amounts of a given raw material needed for each kind of house. Clearly, a matrix is a very succinct way of summarising this information.

Suppose now that the contractor wishes to compute how much of each raw material to obtain in order to fulfil his contracts. Let us denote the matrix above by R; then he would like to obtain something like the product xR, and he would like the product to tell him what orders to make out. The product should have the following form:

$$xR = (5,\ 7,\ 12)\begin{pmatrix} 5 & 20 & 16 & 7 & 17 \\ 7 & 18 & 12 & 9 & 21 \\ 6 & 25 & 8 & 5 & 13 \end{pmatrix}$$

$$= (5.5 + 7.7 + 12.6, 5.20 + 7.18 + 12.25,$$

$$5.16 + 7.12 + 12.8,\ 5.7 + 7.9 + 12.5,$$

$$5.17 + 7.21 + 12.13)$$

$$= (146,\ 526,\ 260,\ 158,\ 388).$$

Thus, we see that the contractor should order 146 units of steel, 526 units of wood, 260 units of glass, 158 units of paint, and 388 units of labour. Observe that the answer we get is a five-component row vector and that each entry in this vector is obtained by taking the vector product of x times the corresponding column of the matrix R.

The contractor is also interested in the prices that he will have to pay for these materials. Suppose that steel costs \$15 per unit, wood costs \$8 per unit, glass costs \$5 per unit, paint costs \$1 per unit, and labour costs \$10 per unit. Then we can write the cost as a column vector as follows:

$$y = \begin{pmatrix} 15 \\ 8 \\ 5 \\ 1 \\ 10 \end{pmatrix}.$$

Here, the product Ry should give the costs of each type of house, so that the multiplication should have the form

$$Ry = \begin{pmatrix} 5 & 20 & 16 & 7 & 17 \\ 7 & 18 & 12 & 9 & 21 \\ 6 & 25 & 8 & 5 & 13 \end{pmatrix} \begin{pmatrix} 15 \\ 8 \\ 5 \\ 1 \\ 10 \end{pmatrix}$$

$$= \begin{pmatrix} 5 \cdot 15 + 20 \cdot 8 + 16 \cdot 5 + 7 \cdot 1 + 17 \cdot 10 \\ 7 \cdot 15 + 18 \cdot 8 + 12 \cdot 5 + 9 \cdot 1 + 21 \cdot 10 \\ 6 \cdot 15 + 25 \cdot 8 + 8 \cdot 5 + 5 \cdot 1 + 13 \cdot 10 \end{pmatrix}$$

$$= \begin{pmatrix} 492 \\ 528 \\ 465 \end{pmatrix}.$$

Thus, the cost of materials for the ranch style house is \$ 492, for the Cape Cod house is \$528, and for the Colonial house \$465.

The final question which the contractor might ask is what is the total cost of raw materials for all the houses he will build. It is easy to see that this is given by the vector xRy. We can find it in two ways as shown below:

$$xRy = (xR)y = (146, 526, 260, 158, 388).\begin{pmatrix} 15 \\ 8 \\ 5 \\ 1 \\ 10 \end{pmatrix} = 11{,}736$$

$$xRy = x(Ry) = (5, 7, 12).\begin{pmatrix} 492 \\ 528 \\ 465 \end{pmatrix} = 11{,}736.$$

The total cost is then \$11,736.

We shall adopt, in general, the above definitions for the multiplication of a matrix times a row or a column vector.

Definition: Let A be an $m \times n$ matrix, let x be an m-component row vector, and let u be an n-component column vector; then we define the products xA and Au as follows:

$$xA = (x_1, x_2, \cdots, x_m)\begin{pmatrix} a_{11} & a_{12} & \cdots & a_{1n} \\ a_{21} & a_{22} & \cdots & a_{2n} \\ & & \cdots & \\ a_{m1} & a_{m2} & \cdots & a_{mn} \end{pmatrix}$$

$$= (x_1a_{11} + x_2a_{21} + \cdots + x_ma_{m1}, x_1a_{12} + x_2a_{22}$$

$$+ \cdots + x_ma_{m2}, \cdots, x_1a_{1n} + x_2a_{2n} + \cdots + x_ma_{mn});$$

$$Au = \begin{pmatrix} a_{11} & a_{12} & \cdots & a_{1n} \\ a_{21} & a_{22} & \cdots & a_{2n} \\ & & \cdots & \\ a_{m1} & a_{m2} & \cdots & a_{mn} \end{pmatrix}\begin{pmatrix} u_1 \\ u_2 \\ \vdots \\ u_n \end{pmatrix}$$

$$= \begin{pmatrix} a_{11}u_1 + a_{12}u_2 + \cdots + a_{1n}u_n \\ a_{21}u_1 + a_{22}u_2 + \cdots + a_{2n}u_n \\ \cdots \\ a_{m1}u_1 + a_{m2}u_2 + \cdots + a_{mn}u_n \end{pmatrix}.$$

The reader will find these formulas easy to work with if he observes that each entry in the products xA or Au is obtained by vector multiplication of x or u by a column or row of the matrix A. Notice that in order to multiply a row vector times a matrix, the number of rows of the matrix must equal the number of components of the vector m and the result is another row vector; similarly, to multiply a matrix times a column vector, the number of columns of the matrix must equal the number of components of the vector, and the result of such a multiplication is another column vector.

Some numerical examples of the multiplication of vectors and matrices are:

$$(1, 0, -1)\begin{pmatrix} 3 & 1 \\ 2 & 3 \\ 2 & 8 \end{pmatrix} = (1.3 + 0.2 - 1.2, +1.1 + 0.3 - 1.8)$$

$$= (1, -7);$$

$$\begin{pmatrix} 3 & 1 & 2 \\ 2 & 3 & 8 \end{pmatrix}\begin{pmatrix} 1 \\ -1 \\ 2 \end{pmatrix} = \begin{pmatrix} 3-1+4 \\ 2-3+16 \end{pmatrix} = \begin{pmatrix} 6 \\ 15 \end{pmatrix};$$

$$\begin{pmatrix} 3 & 2 & -1 \\ 1 & 0 & 2 \\ 0 & 3 & 1 \\ 5 & -4 & 7 \\ -3 & 2 & -1 \end{pmatrix}\begin{pmatrix} 1 \\ 0 \\ -2 \end{pmatrix} = \begin{pmatrix} 5 \\ -3 \\ -2 \\ -9 \\ -1 \end{pmatrix}$$

Observe that if x is an m-component row vector and A is $m \times n$, then xA is an n-component row vector; similarly, if u is an n component column vector, then Au is an m-component column vector. These facts can be observed in the examples above.

Example 2: Consider a Markov chain with transition matrix

$$P = \begin{pmatrix} \frac{1}{3} & \frac{2}{3} \\ \frac{1}{2} & \frac{1}{2} \end{pmatrix}.$$

Choose the initial state by a random device that selects a_1 and a_2 each with probability $\frac{1}{2}$. Let us indicate the choice of initial state by the vector $p^{(0)} = \left(\frac{1}{2}, \frac{1}{2}\right)$ where the first component gives the probability of choosing state a_1 and the second the probability of choosing state a_2. Let us compute the product $p^{(0)}P$. We have

$$p^{(0)}P =$$

$$\left(\frac{1}{2}, \frac{1}{2}\right)\begin{pmatrix} \frac{1}{3} & \frac{2}{3} \\ \frac{1}{2} & \frac{1}{2} \end{pmatrix} = \left(\frac{1}{6} + \frac{1}{4}, \frac{1}{3} + \frac{1}{4}\right) = \left(\frac{5}{12}, \frac{7}{12}\right).$$

One can show that after one step there is probability $\frac{5}{12}$ that the process will be in state a_1 and probability $\frac{7}{12}$ that it will be in state a_2. Let $p^{(1)}$ be the vector whose first component gives the probability of the process being in state a_1 after one step and whose second component gives the probability of it being in state a_2 after one step. In our example we have $p^{(1)}P = \left(\frac{5}{12}, \frac{7}{12}\right) = p^{(0)}P$.

In general, the formula $p^{(1)} = p^{(0)}P$ helds for any Markov process with transition matrix P and initial probability vector $p(0)$.

Example 3: In Example 1 of 'the product of vectors' assume that Smith has two stores at which he can make his purchases, and let us assume that the prices charged at these two stores are slightly different. Let the price vector at the second store be

$$y = \begin{pmatrix} 5 \\ 5 \\ 10 \\ 4 \\ 6 \end{pmatrix} \begin{matrix} \text{cents per apple} \\ \text{cents per peach} \\ \text{cents per lemon} \\ \text{cents per orange} \\ \text{cents per pear.} \end{matrix}$$

Smith now has the option of buying all his purchases at store 1, all at store 2, or buying just the lower-priced items at the store charging the lower price. To help him decide, we form a price matrix as follows:

$$\begin{matrix} \text{Prices,} & \text{Prices,} & \text{Minimum} \\ \text{Store 1} & \text{Store 2} & \text{Price} \end{matrix}$$

$$P = \begin{pmatrix} 4 & 5 & 4 \\ 6 & 5 & 5 \\ 9 & 10 & 9 \\ 5 & 4 & 4 \\ 7 & 6 & 6 \end{pmatrix}.$$

The first column lists the prices of store 1, the second column lists the price of store 2, and the third column lists the lower of these two prices. To compute Smith's bill under the three possible ways he can make his purchases, we compute the produce xP, as follows:

$$xP = (6,\ 12,\ 3,\ 12,\ 6) = \begin{pmatrix} 4 & 5 & 4 \\ 6 & 5 & 5 \\ 9 & 10 & 9 \\ 5 & 4 & 4 \\ 7 & 6 & 6 \end{pmatrix}$$

$$= (225,\ 204,\ 195).$$

We thus see that if Smith buys only in store 1, his bill will be \$2.25; if he buys only in store 2, his bill will be \$2.04; but if he buys each item in the cheaper of the two stores (apples and lemons in store 1, and the rest in store 2), his bill will be \$1.95.

Exactly what Smith will, or should, do depends upon circumstances. If both stores are equally close to him, he will probably split his purchases and obtain the smallest bill. If store 1 is close and store 2 is very far away, he may buy everything at store 1. If store 2 is closer and store 1 is far enough away so that the 9 cents he would save by splitting his purchases is not worth the travel effort, he may buy everything at store 2.

The problem just cited is an example of a *decision problem.* In such problems it is necessary to choose one of several courses of action, or strategies. For each such course of action or strategy, it is possible to compute the cost or worth of such a strategy. The decision-maker will choose a strategy with maximum worth.

Sometimes the worth of an outcome must be measured in psychological units and we then say that we measure the *utility* of an outcome. For the purposes of this book we shall always assume that the utility of an outcome is measured in monetary units, so that we can compare the worths of two different outcomes to the decision-maker.

Example 4: As a second example of a decision problem, consider the following. An urn contains five red, three green, and one white ball. One ball will be drawn at random, and then payments will be made to holders of three kinds of lottery tickets, A, B, and C, according to the following schedule:

$$\begin{array}{r c} & \begin{array}{ccc}\text{Ticket A} & \text{Ticket B} & \text{Ticket C}\end{array} \\ \begin{array}{r}\text{Red} \\ M = \text{Green} \\ \text{White}\end{array} & \begin{pmatrix} 1 & 3 & 0 \\ 4 & 1 & 0 \\ 0 & 0 & 16 \end{pmatrix}. \end{array}$$

Thus, if a red ball is selected, holders of ticket A will get \$1, holders of ticket B will get \$3, and holders of ticket C will get nothing. If green is chosen, the payments are 4, 1, and 0, respectively. If white is chosen, holders of ticket C get \$16, and the others nothing. Which ticket would we prefer to have?

Our decision will depend upon the concept of expected value discussed earlier. The statements "draw a red ball," "draw a green ball," and "draw a white ball" have probabilities $\frac{5}{9}, \frac{3}{9}$, and $\frac{1}{9}$, respectively.

From these probabilities we can calculate the expected value of holding each of the lottery tickets. However, a compact way of performing all these calculations is to compute the product pM, where p is the probability vector

$$p = \left(\tfrac{5}{9}, \tfrac{3}{9}, \tfrac{1}{9}\right)$$

From this we have

$$\begin{aligned} pM &= \left(\tfrac{5}{9}, \tfrac{3}{9}, \tfrac{1}{9}\right)\begin{pmatrix} 1 & 3 & 0 \\ 4 & 1 & 0 \\ 0 & 0 & 16 \end{pmatrix} \\ &= \left(1\cdot\tfrac{5}{9} + 4\cdot\tfrac{3}{9} + 0\cdot\tfrac{1}{9}, 3\cdot\tfrac{5}{9} + 1\cdot\tfrac{3}{9}\right. \\ &\qquad \left. +0\cdot\tfrac{1}{9}, 0\cdot\tfrac{5}{9} + 0\cdot\tfrac{3}{9} + 16\cdot\tfrac{1}{9}\right) \\ &= \left(\tfrac{17}{9}, \tfrac{18}{9}, \tfrac{16}{9}\right). \end{aligned}$$

It is easy to see that the three components of pM give the expected values of holding lottery tickets A, B, and C, respectively. From these numbers we can see that ticket B is the best, A is the next best, and C is third best.

If we have to pay for the tickets, then the cost of the tickets will determine which is the best buy. If each ticket costs \$3 we would be better off by not buying any ticket, since we would then expect to lose money. If each ticket costs \$1 then we should buy ticket B, since it would give us a net expected gain of \$2 - \$1 = \$1. If the first two tickets cost \$2.10, and the third cost \$1.50, we should buy ticket C since it is the only one for which we would have a positive net expectation.

Exercises

1. Perform the following multiplications:

(a) $\begin{pmatrix} 7 & 6 \\ -3 & 2 \end{pmatrix}\begin{pmatrix} 1 \\ -3 \end{pmatrix} = ?$ [*Ans.* $\begin{pmatrix} -11 \\ -9 \end{pmatrix}$]

(b) $(2,-2)\begin{pmatrix} -3 & 2 \\ -1 & 0 \end{pmatrix} = ?$

(c) $\begin{pmatrix} 0 & 3 & 1 \\ 3 & -1 & 7 \\ -5 & 14 & -8 \\ 7 & 2 & 9 \\ 0 & -6 & 10 \end{pmatrix}\begin{pmatrix} 6 \\ 1 \\ -1 \end{pmatrix} = ?$

(d) $(5, 5)\begin{pmatrix} 1 & -1 \\ -1 & 1 \end{pmatrix} = ?$ [Ans. (0,0).]

(e) $\begin{pmatrix} 1 & -1 \\ -1 & 1 \end{pmatrix}\begin{pmatrix} 12 \\ 12 \end{pmatrix} = ?$

(f) $(0, 1, -5)\begin{pmatrix} 3 & 1 & 2 & 0 & -8 \\ 6 & 8 & 2 & 1 & 14 \\ 2 & 15 & 2 & 0 & -5 \end{pmatrix} = ?$

(g) $(x_1, x_2)\begin{pmatrix} a & b \\ c & d \end{pmatrix} = ?$ [Ans. $ax_1 + cx_2$, $bx_1 + dx_2$.]

(h) $\begin{pmatrix} w & x \\ y & z \end{pmatrix}\begin{pmatrix} p_1 \\ p_2 \end{pmatrix} = ?$

(i) $\begin{pmatrix} 1 & 0 & 0 & 0 \\ 0 & 1 & 0 & 0 \\ 0 & 0 & 1 & 0 \\ 0 & 0 & 0 & 1 \end{pmatrix}\begin{pmatrix} u_1 \\ u_2 \\ u_3 \\ u_4 \end{pmatrix} = ?$

(j) $(x_1, x_2, x_3, x_4)\begin{pmatrix} 1 & 0 & 0 & 0 \\ 0 & 1 & 0 & 0 \\ 0 & 0 & 1 & 0 \\ 0 & 0 & 0 & 1 \end{pmatrix} = ?$

2. What number does the matrix in parts (i) and (j) of Exercise 1 resemble?
3. Notice that in Exercise 1 (d), above, the product of a row vector none of whose components is zero and a matrix none of whose components is zero is the zero row vector. Find a second example,

this time using a 3 × 3 matrix, which is similar to this one. Answer the analogous question for Exercise l(e).

4. Consider the matrices

$$A = \begin{pmatrix} a_{11} & a_{12} \\ a_{21} & a_{22} \end{pmatrix}, \; x = \begin{pmatrix} x_1 \\ x_2 \end{pmatrix}, \; b = \begin{pmatrix} b_1 \\ b_2 \end{pmatrix}.$$

(a) Show that the equation $Ax = b$ represents two simultaneous equations in two unknowns.

(b) Show that every set of two simultaneous equations in two unknowns can be written in this form for the proper choice of A and b.

5. When possible, solve for the indicated quantities.

(a) $(x_1, x_2)\begin{pmatrix} 6 & 8 \\ -9 & 0 \end{pmatrix} = (-45, 72)$. Find the vector x.

(b) $(6, 9)\begin{pmatrix} a & b \\ c & d \end{pmatrix} = (12, -15)$. Find the matrix $\begin{pmatrix} a & b \\ c & d \end{pmatrix}$.

In this case can you find more than one solution?

(c) $\begin{pmatrix} 5 & -5 \\ -5 & 5 \end{pmatrix}\begin{pmatrix} u_1 \\ u_2 \end{pmatrix} = \begin{pmatrix} 7 \\ 0 \end{pmatrix}$. Find the vector u.

(d) $\begin{pmatrix} 3 & -7 \\ -12 & 28 \end{pmatrix}\begin{pmatrix} u_1 \\ u_2 \end{pmatrix} = \begin{pmatrix} -1 \\ 4 \end{pmatrix}$. Find u. How many solutions you find?

[*Ans.* Infinitely many solutions, all of the form $u = \begin{pmatrix} k \\ \dfrac{3k+1}{7} \end{pmatrix}$]

6. Solve for the indicated quantities below and give an interpretation for each.

(a) $(1, -1)\begin{pmatrix} 0 & 2 \\ -2 & 4 \end{pmatrix} = a(1, -1)$; find a. [*Ans.* $a = 2$]

(b) $\begin{pmatrix} 1 & 2 \\ 2 & 4 \end{pmatrix}\begin{pmatrix} u_1 \\ u_2 \end{pmatrix} = 5\begin{pmatrix} u_1 \\ u_2 \end{pmatrix}$; find u. How many answers can you find?

[*Ans.* $u = \begin{pmatrix} k \\ 2k \end{pmatrix}$ for any number k]

(c) $\begin{pmatrix} 3 & 3 \\ 6 & 10 \end{pmatrix}\begin{pmatrix} u_1 \\ u_2 \end{pmatrix} = \begin{pmatrix} u_1 \\ u_2 \end{pmatrix}$; find u. How many answers are there?

7. In Example 1 of this section, assume that the contractor is to build eight ranch-style, four Cape Cod, and four colonial-type houses. Recompute, using matrix multiplication, the total cost of raw materials, in two different ways, as in the example.

8. In Example 2 use tree measures to show that $p^{(2)} = p^{(1)}P$.

9. In Example 2 of this section, assume that the initial probability vector is $p^{(0)} = \left(\frac{1}{6}, \frac{5}{6}\right)$. Find the vector $p^{(1)}$. [*Ans.* $\left(\frac{17}{36}, \frac{19}{36}\right)$]

10. Consider the Markov chain with two states whose transition matrix is

$$P = \begin{pmatrix} a & 1-a \\ 1-b & b \end{pmatrix},$$

where a and b are non-negative numbers less than 1. Suppose the initial probability vector for the process is $p^{(0)} = \left(p_1^{(0)}, p_2^{(0)}\right)$, where $p_1^{(0)}$ is the initial probability of choosing state 1 and $p_2^{(0)}$ is the initial probability of choosing state 2. Derive the formulas for the components of the vector $p^{(1)}$.

[*Ans.* $p^{(1)} = ap_1^{(0)} + (1-b)\,p_2^{(0)}, (1-a)\,p_1^{(0)} + bp_2^{(0)}$.]

11. Suppose that $\begin{pmatrix} a & b & c \\ d & e & f \\ g & h & i \end{pmatrix}$ is the transition matrix of a three-state Markov chain. Find $(x, y, z)\begin{pmatrix} a & b & c \\ d & e & f \\ g & h & i \end{pmatrix}\begin{pmatrix} 1 \\ 1 \\ 1 \end{pmatrix}$.

[*Ans.* $x + y + z$.]

12. The following matrix gives the vitamin contents of three food items, in conveniently chosen units:

$$\begin{array}{lcccc} \text{Vitamin:} & A & B & C & D \\ \text{Food I:} & .5 & .5 & 0 & 0 \\ \text{Food II:} & .3 & 0 & .2 & .1 \\ \text{Food III:} & .1 & .1 & .2 & .5 \end{array}$$

If we eat 11 units of food I, 6 units of food II, and 4 units of food III, how much of each type of vitamin have we consumed? If we pay only for the vitamin content of each food, paying 10 cents, 20 cents, 25 cents, and 50 cents, respectively, for units of the four vitamins, how much does a unit of each type of food cost? Compute in two ways the total cost of the food we ate.[*Partial Ans.* $3.75]

13. In Example 3, by how much would store 1 have to reduce the price of apples to make Smith's total purchases less expensive there than at store 2?

14. In Example 3, find the store at which the total cost to Smith is the least when he wishes to purchase

(a) $x = (4, 1, 2, 0, 1)$. [*Ans.* Store 1; cost 47 cents.]

(b) $x = (1, 3, 2, 4, 0)$.

(c) $x = (2, 2, 2, 0, 2)$.

15. In Example 4, let us assume that an individual chooses ticket 1 with probability r_1, ticket 2 with probability r_2, and ticket 3 with probability r_3. Let $r = \begin{pmatrix} r_1 \\ r_2 \\ r_3 \end{pmatrix}$. Give an interpretation for pMr. Compute this for $r_1 = \frac{1}{4}$, $r_2 = \frac{1}{4}$, and $r_3 = \frac{1}{2}$.

[*Ans.* $pMr = \frac{67}{36}$, which is the expected return.]

16. A game room contains three pinball machines. Either a game on one of these machines terminates normally, or else the machine refunds enough money for one or two games. A game also ends if the machine is tilted. The probability of each of these events is given by the following matrix:

$$M = \begin{array}{l} \\ \text{Machine 1} \\ \text{Machine 2} \\ \text{Machine 3} \end{array} \begin{array}{cccc} \text{Normal} & \text{Refunds I Game} & \text{Refunds 2} & \text{Tilt} \\ \end{array}$$

$$M = \begin{array}{l} \text{Machine 1} \\ \text{Machine 2} \\ \text{Machine 3} \end{array} \begin{pmatrix} .8 & .09 & .01 & .1 \\ .75 & .045 & .005 & .2 \\ .9 & .04 & .01 & .05 \end{pmatrix}.$$

Assume that n = (25, 20, 30) represents the number of games played on machines 1, 2, and 3, respectively. Compute and interpret nM.

17. In Exercise 16, Suppose it costs $10 to play one game.
    (a) Construct a column vector with entries being the profit per game made by the owner of the machines for each of the different outcomes.
    (b) What is the expected profit made by the owner if 100 people play machine 1, 80 play machine 2, and 120 play machine 3?
    (c) Due to space limitations the owner must sell one of the machines. Which one should he sell?

18. (a) Consider the matrices

$$P = \begin{pmatrix} \frac{1}{3} & \frac{2}{3} \\ \frac{3}{4} & \frac{1}{4} \end{pmatrix} \text{ and } f = \begin{pmatrix} 1 \\ 1 \end{pmatrix}.$$

Show that $Pf = f$. The vector f is called a *fixed vector* on the right of P.

(b) Let $w = \left(\frac{9}{17}, \frac{8}{17}\right)$ and let P be the matrix in part (a). Show that $wP = w$. For this reason w is called a *fixed vector* on the left of P.

19. Let P be the matrix of transition probabilities for a Markov chain having n states, and let f be a column matrix all of whose entries are l's. Show that $Pf = f$. [*Hint:* Exercise 18 provides a special case.]

20. Let A, B, and C be matrices of the same shape, and let h and k be numbers. Use the ordinary rules for numbers plus the definitions of this section to show that the following laws hold:
    (a1) $A + B = B + A$ (commutative law of addition).
    (a2) $A + (B + C) = (A + B) + C$ (Associative Law of addition).
    (a3) If 0 is the zero matrix of the same shape, then $A + 0 = A$ (additive identity law).
    (a4) Define $-A = (-1)A$; *then* $A + (-A) = 0$ (additive inverse law).
    (s1) $h(kA) = (hk)A$ (mixed Associative Law).
    (s2) $1A = A$ for all A (unity law).

(s3) $h(A + B) = hA + hB$ (first distributive law).

(s4) $(h + k)A = hA + kA$ (second distributive law).

21. A company is considering which of three methods of production it should use in producing three goods, A, B, and C. The amount of each good produced by each method is shown in the matrix:

$$\begin{array}{c} \quad\; A \;\; B \;\; C \\ R = \begin{pmatrix} 2 & 3 & 1 \\ 1 & 2 & 3 \\ 2 & 4 & 1 \end{pmatrix} \begin{array}{l} \text{Method 1} \\ \text{Method 2} \\ \text{Method 3.} \end{array} \end{array}$$

Let p be a vector whose components represent the profit per unit for each of the goods. What does the vector Rp represent? Find three different vectors p such that under each of these profit vectors a different method would be most profitable.[*Partial Ans.* For $p = \begin{pmatrix} 10 \\ 8 \\ 7 \end{pmatrix}$ method 3 is most profitable.]

Addition and Multiplication of Matrices

Two matrices of the same shape—that is, having the same number of rows and columns—can be added together by adding corresponding components. For example, if A and B are two 2×3 matrices, we have

$$A + B = \begin{pmatrix} a_{11} & a_{12} & a_{13} \\ a_{21} & a_{22} & a_{23} \end{pmatrix} + \begin{pmatrix} b_{11} & b_{12} & b_{13} \\ b_{21} & b_{22} & b_{23} \end{pmatrix}$$

$$= \begin{pmatrix} a_{11} + b_{11} & a_{12} + b_{12} & a_{13} + b_{13} \\ a_{21} + b_{21} & a_{22} + b_{22} & a_{23} + b_{23} \end{pmatrix}.$$

Observe that the addition of vectors (row or column) is simply a special case of the addition of matrices. Numerical examples of the addition of matrices are

$$(1, 0, -2) + (0, 5, 0) = (1, 5, -2);$$

$$\begin{pmatrix} 1 & 0 \\ 0 & 1 \end{pmatrix} + \begin{pmatrix} -1 & 0 \\ 0 & -1 \end{pmatrix} = \begin{pmatrix} 0 & 0 \\ 0 & 0 \end{pmatrix};$$

$$\begin{pmatrix} 7 & 0 & 0 \\ -3 & 1 & -6 \\ 4 & 0 & 7 \\ 0 & -2 & -2 \\ 1 & 1 & 1 \end{pmatrix} + \begin{pmatrix} -8 & 0 & 1 \\ 4 & 5 & -1 \\ 0 & 3 & 0 \\ -1 & 1 & -1 \\ 0 & -4 & 2 \end{pmatrix} = \begin{pmatrix} -1 & 0 & 1 \\ 1 & 6 & -7 \\ 4 & 3 & 7 \\ -1 & 1 & -3 \\ 1 & -3 & 3 \end{pmatrix}.$$

Other examples occur in the exercises. The reader should observe that we do not add matrices of different shapes.

If A is a matrix and k is any number, we define the matrix kA as

$$kA = k\begin{pmatrix} a_{11} & a_{12} & \cdots & a_{1n} \\ a_{21} & a_{22} & \cdots & a_{2n} \\ \vdots & \vdots & \cdots & \vdots \\ a_{m1} & a_{m2} & \cdots & a_{mn} \end{pmatrix} = \begin{pmatrix} ka_{11} & ka_{12} & \cdots & ka_{1n} \\ ka_{21} & ka_{22} & \cdots & ka_{2n} \\ \vdots & \vdots & \cdots & \vdots \\ ka_{m1} & ka_{m2} & \cdots & ka_{mn} \end{pmatrix}.$$

Observe that this is merely entry wise multiplication, as was the analogous concept for vectors. Examples of multiplication of matrices by constants are

$$-2\begin{pmatrix} 7 & -2 & 8 \\ 0 & 5 & -1 \end{pmatrix} = \begin{pmatrix} -14 & 4 & -16 \\ 0 & -10 & 2 \end{pmatrix};$$

$$6\begin{pmatrix} 1 & 0 \\ 0 & 1 \\ 3 & -4 \end{pmatrix} = \begin{pmatrix} 6 & 0 \\ 0 & 6 \\ 18 & -24 \end{pmatrix}.$$

The multiplication of a vector by a number is a special case of the multiplication of a matrix by a number.

Under certain conditions two matrices can be multiplied together to give a new matrix. As an example, let A be a 2×3 matrix and B be a 3×2 matrix. Then the product AB is found to be

$$AB = \begin{pmatrix} a_{11} & a_{12} & a_{13} \\ a_{21} & a_{22} & a_{23} \end{pmatrix} \begin{pmatrix} b_{11} & b_{12} \\ b_{21} & b_{22} \\ b_{31} & b_{32} \end{pmatrix}$$

$$= \begin{pmatrix} a_{11}b_{11} + a_{12}b_{21} + a_{13}b_{31} & a_{11}b_{12} + a_{12}b_{22} + a_{13}b_{32} \\ a_{21}b_{11} + a_{22}b_{21} + a_{23}b_{31} & a_{21}b_{12} + a_{22}b_{22} + a_{23}b_{32} \end{pmatrix}.$$

Observe that the product is a 2 × 2 matrix. Also notice that each entry in the new matrix is the product of one of the rows of A times one of the columns of B; for example, the entry in the second row and first column is found as the product

$$\begin{pmatrix} a_{21} & a_{22} & a_{23} \end{pmatrix} \begin{pmatrix} b_{11} \\ b_{21} \\ b_{31} \end{pmatrix} = a_{21}b_{11} + a_{22}b_{21} + a_{23}b_{31}.$$

The following definition holds for the general case of matrix multiplication:

Definition: Let A be an $m \times k$ matrix and B be a $k \times n$ matrix; then the product matrix C = AB is an $m \times n$ matrix whose components are

$$c_{ij} = \begin{pmatrix} a_{i1} & a_{i2} & \cdots & a_{ik} \end{pmatrix} \begin{pmatrix} b_{1j} \\ b_{2j} \\ \vdots \\ b_{kj} \end{pmatrix}$$

$$= a_{i1}b_{1j} + a_{i2}b_{2j} + \cdots + a_{ik}b_{kj}.$$

The important things to remember about this definition are: first, in order to be able to multiply matrix A times matrix B, the number of columns of A must be equal to the number of rows of B; second, the product matrix C = AB has the same number of rows as A and the same number of columns as B; finally, to get the entry in the ith row and jth column of AB we multiply the ith row of A times the jth column of B. Notice that the product of a vector times a matrix is a special case of matrix multiplication.

Below are several examples of matrix multiplication:

$$\begin{pmatrix} 2 & -1 \\ 0 & 3 \end{pmatrix} \begin{pmatrix} 7 & 0 \\ -2 & -3 \end{pmatrix} = \begin{pmatrix} 16 & 3 \\ -6 & -9 \end{pmatrix};$$

$$\begin{pmatrix} 3 & 0 & 1 \\ -1 & 2 & 0 \\ 0 & 0 & 2 \end{pmatrix}\begin{pmatrix} 1 & 0 & 0 \\ 0 & -1 & 0 \\ 1 & 1 & 1 \end{pmatrix} = \begin{pmatrix} 4 & 1 & 1 \\ -1 & -2 & 0 \\ 2 & 2 & 2 \end{pmatrix};$$

$$\begin{pmatrix} 3 & 1 & 4 \\ 2 & 0 & 5 \end{pmatrix}\begin{pmatrix} 1 & 3 & 0 & 0 \\ 1 & 1 & 0 & 0 \\ 0 & 0 & 1 & 1 \end{pmatrix} = \begin{pmatrix} 4 & 10 & 4 & 4 \\ 2 & 6 & 5 & 5 \end{pmatrix}.$$

We next ask how we multiply more than two matrices together. Let A be an $m \times h$ matrix, let B be an $h \times k$ matrix, and let C be a $k \times n$ matrix. Then we can certainly define the products $(AB)C$ and $A(BC)$. It turns out that these two products are equal, and we define the product ABC to be their common value; that is,

$$ABC = A(BC) = (AB)C.$$

The rule expressed in the above equation is called the *associative law* for multiplication. We shall not prove the associative law here, although the student will be asked to check an example of it in Exercise 5.

If A and B are square matrices of the same size, then they can be multiplied in either order. It is not true, however, that the product AB is necessarily equal to the product BA. For example, if

$$A = \begin{pmatrix} 1 & 1 \\ 0 & 0 \end{pmatrix} \text{ and } B = \begin{pmatrix} 1 & 0 \\ 1 & 0 \end{pmatrix},$$

then we have

$$AB = \begin{pmatrix} 1 & 1 \\ 0 & 0 \end{pmatrix}\begin{pmatrix} 1 & 0 \\ 1 & 0 \end{pmatrix} = \begin{pmatrix} 2 & 0 \\ 0 & 0 \end{pmatrix},$$

$$AB = \begin{pmatrix} 1 & 1 \\ 0 & 0 \end{pmatrix}\begin{pmatrix} 1 & 0 \\ 1 & 0 \end{pmatrix} = \begin{pmatrix} 2 & 0 \\ 0 & 0 \end{pmatrix},$$

whereas

$$BA = \begin{pmatrix} 1 & 0 \\ 1 & 0 \end{pmatrix}\begin{pmatrix} 1 & 1 \\ 0 & 0 \end{pmatrix} = \begin{pmatrix} 1 & 1 \\ 1 & 1 \end{pmatrix},$$

and it is clear that $AB \neq BA$.

Exercises

1. **Perform the following matrix operations:**

(a) $\begin{pmatrix} 3 & 2 \\ 1 & 4 \\ 5 & 3 \end{pmatrix} - \begin{pmatrix} 0 & 7 \\ -3 & -3 \\ 8 & 1 \end{pmatrix} = ?$ [*Ans.* $\begin{pmatrix} 3 & -5 \\ 4 & 7 \\ -3 & 2 \end{pmatrix}$]

(b) $3\begin{pmatrix} 1 & 0 & 4 \\ 3 & 1 & -2 \\ 2 & 1 & -1 \end{pmatrix} + \begin{pmatrix} 2 & 0 & 2 \\ 5 & 0 & 1 \\ 3 & 5 & 2 \end{pmatrix} = ?$

(c) $\begin{pmatrix} 1 & 8 \\ -3 & 0 \end{pmatrix}\begin{pmatrix} 2 & 1 \\ -1 & 2 \end{pmatrix} = ?$

(d) $\begin{pmatrix} 0 & 10 \\ 3 & 6 \end{pmatrix}\begin{pmatrix} 8 & 1 & 3 \\ -6 & -1 & 2 \end{pmatrix} = ?$

(e) $\begin{pmatrix} 8 & -6 \\ 1 & -1 \\ 3 & 2 \end{pmatrix}\begin{pmatrix} 0 & 3 \\ 10 & 6 \end{pmatrix} = ?$

(f) $\begin{pmatrix} 1 & 3 \\ 1 & 5 \end{pmatrix}\left[\begin{pmatrix} 3 & 6 \\ 4 & 4 \end{pmatrix} - 8\begin{pmatrix} 5 & 4 \\ 0 & 3 \end{pmatrix}\right] = ?$

(g) $\begin{pmatrix} 1 & -2 \\ 1 & -1 \end{pmatrix}\begin{pmatrix} 2 \\ 1 \end{pmatrix} = ?$

(h) $\begin{pmatrix} 7 & 9 & 2 \\ 4 & 9 & 6 \\ 5 & 6 & 0 \end{pmatrix}\begin{pmatrix} 3 & 3 & 5 \\ 3 & 9 & 4 \\ 4 & 5 & 7 \end{pmatrix} = ?$ [*Ans.* $\begin{pmatrix} 56 & 112 & 85 \\ 63 & 123 & 98 \\ 33 & 69 & 49 \end{pmatrix}$.]

(i) $\left[2\begin{pmatrix} 6 & 0 & 1 \\ 1 & -3 & 2 \end{pmatrix} + \begin{pmatrix} 4 & 0 & -4 \\ 2 & 1 & -1 \end{pmatrix}\right]\begin{pmatrix} 3 & 0 & 1 \\ -1 & 2 & 0 \\ 0 & 0 & 0 \end{pmatrix} = ?$

(j) $\begin{pmatrix} 1 & -1 \\ -1 & 1 \end{pmatrix}\begin{pmatrix} 1 & -1 \\ -1 & 1 \end{pmatrix} = ?$

2. Consider the matrices $A=\begin{pmatrix}-19 & 2\\ 14 & -10\\ 5 & 0\end{pmatrix}$, $B=\begin{pmatrix}7 & 2 & 19\\ 4 & 26 & -2\\ 13 & 0 & 7\end{pmatrix}$,

$C=\begin{pmatrix}4 & 1 & -1\\ 6 & 3 & -5\end{pmatrix}$, and $D=\begin{pmatrix}10 & -7\\ 4 & 15\end{pmatrix}$. Their shapes are 3 × 2, 3 × 3, 2 × 3, and 2 × 2, respectively. What is the shape of:

(a) *AC*?

(b) *CB*?

(c) *DC*? [*Ans.* 2 × 3]

(d) *ACB*?

(e) *BAC*?

(f) *DCB*?

(g) *DCBA*?

(h) *ADCB*? [*Ans.* 3 × 3.]

3. In Exercise 2, find the component:

(a) In the second row and second column of *AC*. [*Ans.* –16.]

(b) In the first row and the third column of *CB*.

(c) In the last row and last column of *AC*.

(d) In the last row and last column of *CA*. [*Ans.* –18.]

(e) In the second row and first column of *DC*.

4. Let *A* be any 3 × 3 matrix and let *I* be the matrix

$$I=\begin{pmatrix}1 & 0 & 0\\ 0 & 1 & 0\\ 0 & 0 & 1\end{pmatrix}.$$

Show that $AI = IA = A$. The matrix *I* acts for the products of matrices in the same way that the number 1 acts for products of numbers. For this reason it is called the *identity matrix*.

5. Verify the Associative Law for the special case when

$$A=\begin{pmatrix}-1 & 3 & 3\\ 6 & 5 & 1\end{pmatrix},\ B=\begin{pmatrix}4 & -3 & 3\\ 8 & 0 & 5\\ 5 & -4 & 3\end{pmatrix},\ \text{and } C=\begin{pmatrix}9 & 4\\ 7 & -4\\ 0 & 1\end{pmatrix}.$$

6. The commutative law for addition is

   $A + B = B + A$

   for any two matrices A and B of the same shape. Prove that the commutative law for addition is true from the definition of matrix addition and from the fact that it is true for ordinary numbers.

7. Show that there is *not* a commutative law for matrix multiplication by finding two 2×2 matrices A and B different than the ones in the text such that $A \cdot B \neq B \cdot A$.

8. The distributive law for numbers and matrices is

   $$k(A + B) = kA + kB$$

   for any number k and any two matrices A and B of the same shape. Prove that this law holds from the definitions of numerical multiplication of matrices, addition of matrices, and the ordinary rules for numbers.

9. The distributive laws for matrices are

   $$(A + B)C = AC + BC,$$
   $$C(A + B) = CA + CB,$$

   where A, B, and C are matrices of suitable shapes. Show that these laws hold from the definitions of matrix multiplication and addition, and the ordinary rules for numbers.

10. Let A be any 3×3 matrix and let 0 be the matrix.

    $$0 = \begin{pmatrix} 0 & 0 & 0 \\ 0 & 0 & 0 \\ 0 & 0 & 0 \end{pmatrix}.$$

    Show that $A0 = 0A = 0$ for any A. Also show that $A + 0 = 0 + A = A$ for any A. The matrix 0 acts for matrices in the same way that the number 0 acts for numbers. For this reason it is called the *zero matrix.*

11. Show that for any square matrix A there is a matrix B of the same shape as A such that $A + B = 0$. (B is called the *additive inverse* of A.)

12. Find the additive inverse of each of the following matrices:

    (a) $\begin{pmatrix} 1 & 1 \\ 2 & -1 \end{pmatrix}$.

(b) $\begin{pmatrix} -3 & 2 \\ 6 & 1 \end{pmatrix}$.

(c) $\begin{pmatrix} -3 & 2 & 4 \\ 0 & 8 & 3 \\ 5 & 4 & -7 \end{pmatrix}$.

13. If $A = \begin{pmatrix} 0 & 0 \\ 0 & 1 \end{pmatrix}$ and $B = \begin{pmatrix} 1 & 0 \\ 0 & 0 \end{pmatrix}$, show that $AB = \begin{pmatrix} 0 & 0 \\ 0 & 0 \end{pmatrix}$. Thus, the product of two matrices can be the zero matrix even though neither of the matrices is itself zero. Find another example that illustrates this point.

14. If A is a square matrix, it can be multiplied by itself; hence we can define (using the associative law)

$A^2 = A . A$

$A^3 = A^2 . A = A . A . A$

$A^n = A^{n-1} . A = A . A ..... A$ (*n* factors).

These are naturally called "powers" of the matrix A, A^2 being called the square of A, A^3 the cube of A, etc.

(a) Compute A^2, A^3, and A^4 for $A = \begin{pmatrix} 2 & 0 \\ 3 & -1 \end{pmatrix}$.

(b) If I and 0 are the matrices defined in Exercises 4 and 10, find I^2, I^3, I^n, 0^2, 0^3, and 0^n.

(c) If $A = \begin{pmatrix} 1 & 1 \\ 1 & 1 \end{pmatrix}$, find A^n.

(d) If $A = \begin{pmatrix} 0 & 1 & 2 \\ 0 & 0 & -1 \\ 0 & 0 & 0 \end{pmatrix}$, find A^2, A^3, and A^n.

15. Find the matrices $P^{(2)}$ and $P^{(3)}$ for the Markov chain whose transition matrix is $P = \begin{pmatrix} \frac{1}{2} & \frac{1}{2} \\ \frac{2}{3} & \frac{1}{3} \end{pmatrix}$. Compute P^2 and P^3 and compare the results.

16. Cube the matrix

$$\begin{pmatrix} 0 & 1 & 0 \\ 0 & \frac{1}{2} & \frac{1}{2} \\ \frac{1}{3} & 0 & \frac{2}{3} \end{pmatrix}.$$

17. Consider a two-state Markov process whose transition matrix is

$$P = \begin{pmatrix} p_{11} & p_{12} \\ p_{21} & p_{22} \end{pmatrix}.$$

(a) Assuming that the process starts in state 1, draw the tree and set up tree measures for three stages of the process. Do the same, assuming that the process starts in state 2.

(b) Using the trees drawn in (a), compute the quantities $p_{11}^{(3)}, p_{12}^{(3)}, p_{21}^{(3)}, p_{22}^{(3)}$. Write the matrix $P^{(3)}$.

(c) Compute the cube $P^{(3)}$ of the matrix P.

(d) Compare the answers you found in parts (b) and (c) and show that $P^{(3)} = P^3$.

18. Let $A = \begin{pmatrix} 1 & 0 \\ 1 & 2 \end{pmatrix}$.

(a) Find a matrix B such that $AB = \begin{pmatrix} 1 & 0 \\ 0 & 1 \end{pmatrix}$. Show that $BA = \begin{pmatrix} 1 & 0 \\ 0 & 1 \end{pmatrix}$ as well.

(b) Find a matrix C such that $AC = \begin{pmatrix} 1 & 1 \\ 1 & 3 \end{pmatrix}$. What is CA?

19. A diagonal matrix is square and its only non-zero entries are on the main diagonal. For instance, the matrices

$$A = \begin{pmatrix} 1 & 0 \\ 0 & 4 \end{pmatrix}, \quad B = \begin{pmatrix} 3 & 0 \\ 0 & 2 \end{pmatrix}$$

are 2×2 diagonal matrices.

(a) Show that A and B commute, i.e., $AB = BA$.

(b) Show that any pair of diagonal matrices of the same size commute when multiplied together.

20. A *scalar matrix* is a diagonal matrix in which all the entries on the main diagonal are equal. For instance, the matrices $A = \begin{pmatrix} 3 & 0 \\ 0 & 3 \end{pmatrix}$ and $B = \begin{pmatrix} -7 & 0 \\ 0 & -7 \end{pmatrix}$ are 2 × 2 scalar matrices.
    (a) Show that A commutes with *any* 2 × 2 matrix.
    (b) Show that any scalar matrix A can be written as kI, where k is a number and I is the identity matrix of the appropriate size.
    (c) Show that any scalar matrix commutes with any other matrix of the same size.

21. In Example 1 (of 'Matrices and their combination with vectors') assume that the contractor wishes to take into account the cost of transporting raw materials to the building site as well as the purchasing cost. Suppose the costs are as given in the matrix below:

$$\begin{matrix} & \text{Purchase} & \text{Transport} & \\ Q = & \begin{pmatrix} 15 & 4.5 \\ 8 & 2 \\ 5 & 2 \\ 1 & 0.5 \\ 10 & 0 \end{pmatrix} & \begin{matrix} \text{Steel} \\ \text{Wood} \\ \text{Glass} \\ \text{Paint} \\ \text{Labor} \end{matrix} \end{matrix}$$

Referring to the example:
    (a) By computing the product RQ find a 3 × 2 matrix whose entries give the purchase and transportation costs of the materials for each kind of house.
    (b) Find the product xRQ, which is a two-component row vector whose first component gives the total purchase price and second component gives the total transportation cost.
    (c) Let $z = \begin{pmatrix} 1 \\ 1 \end{pmatrix}$ and then compute $xRQz$, which is a number giving the total cost of materials and transportation for all the houses being built. [*Ans.* 14,304]

22. A candy company packages four sizes of assorted chocolates: the Sampler, Sweetheart, Matinee, and Jumbo boxes. The Sampler

contains 3 almond creams, 4 chocolate nougats, 5 caramel creams, and 3 nut clusters. The Sweetheart contains 6 almond creams, 4 chocolate nougats, 8 caramel creams, and 7 nut clusters. In the Matinee box are 10 almond creams, 15 chocolate nougats, 5 caramel creams, and 5 nut clusters. The Jumbo assortment has 10 almond creams, 15 chocolate nougats, 15 caramel creams, and 10 nut clusters. The company uses as ingredients in its manufacturing process chocolate, nuts, and cream filling. An almond cream contains 1 unit of chocolate, 2 units of nuts, and 2 units of cream filling; a chocolate nougat, 2 units each of chocolate and cream filling; a caramel cream, 4 units of cream filling; and a nut cluster, 3 units of nuts and 2 units of chocolate. Suppose a unit of chocolate costs ¢1, a unit of nuts ¢1.6, and a unit of cream filling ¢2. If the company packages 50 Samplers, 75 Sweethearts, 40 Matinees, and 100 Jumbos,

(a) What is the total number of each type of candy produced?

(b) What is the total number of units of each ingredient used to make all the candy?

(c) What is the total cost of the candy in each box?

(d) What is the total cost of all the candy in all the boxes?

[*Ans.* \$651.50]

Solution of Linear Equations

There are many occasions when the simultaneous solutions of Linear Equations is important. In this section we shall develop methods for finding out whether a set of Linear Equations has solutions, and for finding all such solutions.

Example 1: Consider the following example of three Linear Equations in three unknowns:

$$x_1 + 4x_2 + 3x_3 = 1 \tag{1}$$

$$2x_1 + 5x_2 + 4x_3 = 4 \tag{2}$$

$$x_1 - 3x_2 - 2x_3 = 5 \tag{3}$$

Equations such as these, containing one or more variables, are called *open statements*. Statement (1) is true for some values of the variables (for instance, when $x_x = 1$, $x_2 = 0$, and $x_3 = 0$), and false for other values of the variables (for instance, when $x_1 = 0, x_2 = 1$, and $x_3 = 0$). The truth set of (1) is the set of all vectors $\begin{pmatrix} x_1 \\ x_2 \\ x_3 \end{pmatrix}$ for which (1) is true. Similarly, the truth set of the three simultaneous equations (1), (2), and (3) is the set of all vectors $\begin{pmatrix} x_1 \\ x_2 \\ x_3 \end{pmatrix}$ which make true their conjunction

$$(x_1 + 4x_2 + 3x_3 = 1) \wedge (2x_1 + 5x_2 + 4x_3 = 4) \wedge (x_1 - 3x_2 - 2x_3 = 5).$$

When we say that we solve a set of simultaneous equations, we mean that we determine the truth set of their conjunction.

Before we discuss the solution of these equations we note that they can be written as a single equation in matrix form as follows:

$$\begin{pmatrix} 1 & 4 & 3 \\ 2 & 5 & 4 \\ 1 & -3 & -2 \end{pmatrix} \begin{pmatrix} x_1 \\ x_2 \\ x_3 \end{pmatrix} = \begin{pmatrix} 1 \\ 4 \\ 5 \end{pmatrix}.$$

One of the uses of vector and matrix notation is in writing a large number of linear equations in a single simple matrix equation such as the one above. It also leads to the detached coefficient form of solving simultaneous equations that we shall discuss at the end of the present section and in the next section.

The method of solving the linear equations above is the following. First we use equation (1) to eliminate the variable x_1 from equations (2) and (3); i.e., we subtract 2 times (1) from (2) and then subtract (1) from (3), giving

$$x_1 + 4x_2 + 3x_3 = 1 \qquad (1')$$

$$-3x_2 - 2x_3 = 2 \qquad (2')$$

$$-7x_2 - 5x_3 = 4 \qquad (3')$$

By *pivoting* we shall mean the operation of using an equation to eliminate a variable from the other equations. The *pivot* is the coefficient of the variable being eliminated. In this case the pivot is 1. Next we pivot on − 3 in (2'): divide equation (2') through by the coefficient of x_2, namely,−3, obtaining $x_2 + \frac{2}{3}x_3 = -\frac{2}{3}$. We use this equation to eliminate

x_2 from each of the other two equations. In order to do this we subtract 4 times this equation from (1') and add 7 times this equation to (3'), obtaining

$$x_1 + 0 + \tfrac{1}{3}x_3 = \tfrac{11}{3} \qquad (1'')$$

$$x_2 + \tfrac{2}{3}x_3 = \tfrac{2}{3} \qquad (2'')$$

$$-\tfrac{1}{3}x_3 = -\tfrac{2}{3} \qquad (3'')$$

The last step is to pivot on $-\frac{1}{3}$ by dividing through (3") by $-\frac{1}{3}$, which is the coefficient of x_3, obtaining the equation $x_3 = 2$; we use this equation to eliminate x_3 from the first two equations as follows:

$$x_1 + 0 + 0 = 3 \qquad (1''')$$

$$x_2 + 0 = -2 \qquad (2''')$$

$$x_3 = 2 \qquad (3''')$$

The solution can now be read from these equations as $x_1 = 3$, $x_2 = -2$, and $x_3 = 2$. The reader should substitute these values into the original equations (1), (2), and (3) above to see that the solution has actually been obtained.

In the example just discussed we saw that there was only one solution to the set of three simultaneous equations in three variables. Example 2 will be one in which there is more than one solution, and Example 3 will be one in which there are no solutions to a set of three simultaneous equations in three variables.

Example 2: Consider the following linear equations:

$$x_1 - 2x_2 - 3x_3 = 2 \qquad (4)$$

$$x_1 - 4x_2 - 13x_3 = 14 \qquad (5)$$

$$-3x_1 + 5x_2 + 4x_3 = 0 \qquad (6)$$

Let us proceed as before and use equation (4) to eliminate the variable x_1 from the other two equations. Pivoting on the 1 coefficient of x_1 in (4), we have

$$x_1 - 2x_2 - 3x_3 = 2 \qquad (4')$$

$$-2x_2 - 10x_3 = 12 \qquad (5')$$

$$-x_2 - 5x_3 = 6 \qquad (6')$$

Proceeding as before, we divide equation (5') by –2, obtaining the equation $x_2 + 5x_3 = -6$. We use this equation to eliminate the variable x_2 from each of the other equations—namely, we add twice this equation to (4') and then add the equation to (6'):

$$x_1 + 0 + 7x_3 = -10 \tag{4''}$$

$$x_2 + 5x_3 = -6 \tag{5''}$$

$$0 = 0 \tag{6''}$$

Observe that we have eliminated the last equation completely! We also see that the variable x_3 can be chosen completely arbitrarily in these equations. To emphasise this, we move the terms involving x_3 to the right-hand side, giving

$$x_1 = -10 - 7x_3 \tag{4'''}$$

$$x_2 = -6 - 5x_3 \tag{5'''}$$

The reader should check, by substituting these values of x_1 and x_2 into equations (4), (5), and (6), that they are solutions regardless of the value of x_3. Let us also substitute particular values for x_3 to obtain numerical solutions. Thus, if we let $x_3 = 1, 0, -2$, respectively, and compute the resulting numbers, using (4''') and (5'''), we obtain the following numerical solutions:

$$x_1 = -17, \quad x_2 = -11, \quad x_3 = 1$$

$$x_1 = -10, \quad x_2 = -6, \quad x_3 = 0$$

$$x_1 = 4, \quad x_2 = 4, \quad x_3 = -2.$$

The reader should also substitute these numbers into (4), (5), and (6) to show that they are solutions. To summarise, our second example has an infinite number of solutions, one for each numerical value of x_3 which is substituted into equations (4''') and (5''').

Example 3: Suppose that we modify equation (6) by changing the number on the right-hand side to 2. Then we have

$$x_1 - 2x_2 - 3x = 2 \tag{7}$$

$$x_1 - 4x_2 - 13x_3 = 14 \tag{8}$$

$$-3x_1 + 5x_2 + 4x_3 = 2 \tag{9}$$

If we carry out the same procedure as before and use (7) to eliminate x_1 from (8) and (9), we obtain

$$x_1 - 2x_2 - 3x_3 = 2 \tag{7'}$$

$$-2x_2 - 10x_3 = 12 \tag{8'}$$

$$-x_2 - 5x_3 = 8 \tag{9'}$$

We divide (8') by –2, the coefficient of x_2, obtaining, as before, $x_2 + 5x_3 = -6$. Using this equation to eliminate x_2 from the other two equations, we have

$$x_1 + 0 + 7x_3 = -10 \tag{7''}$$

$$x_2 + 5x_3 = -6 \tag{8''}$$

$$0 = 2 \tag{9''}$$

Observe that the last equation is *logically false,* that is, false for all values of x_1, x_2, x_3. Because our elimination procedure has led to a false result we conclude that the equations (7), (8), and (9) have no solution. The student should always keep in mind that this possibility exists when considering simultaneous equations.

In the examples above the equations we considered had the same number of variables as equations. The next example has more variables than equations and the last has more equations than variables.

Example 4: Consider the following two equations in three variables:

$$-4x_1 + 3x_2 + 2x_3 = -2 \tag{10}$$

$$5x_1 - 4x_2 + x_3 = 3 \tag{11}$$

Using the elimination method outlined above, we divide (10) by –4, and then subtract 5 times the result from (11), obtaining

$$x_1 - \tfrac{3}{4}x_2 - \tfrac{1}{2}x_3 = \tfrac{1}{2} \tag{10'}$$

$$-\tfrac{1}{4}x_2 + \tfrac{7}{2}x_3 = \tfrac{1}{2} \tag{11'}$$

Multiplying (11') by –4 and using it to eliminate x_2 from (10'), we have

$$x_1 + 0 - 11x_3 = -1 \qquad (10'')$$

$$x_2 - 14x_3 = -2 \qquad (11'')$$

We can now let x_3 take on any value whatsoever and solve these equations for x_1 and x_2. We emphasise this fact by rewriting them as in Example 2 as

$$x_1 = 11x_3 - 1 \qquad (10''')$$

$$x_2 = 14x_3 - 2 \qquad (11''')$$

The reader should check that these are solutions and also, by choosing specific values for x_3, find numerical solutions to these equations.

Example 5: Let us consider the other possibility suggested by Example 4, namely, the case in which we have more equations than variables. Consider the following equations:

$$-4x_1 + 3x_2 = 2 \qquad (12)$$

$$5x_1 - 4x_2 = 0 \qquad (13)$$

$$2x_1 - x_2 = a, \qquad (14)$$

where a is an arbitrary number. Using equation (12) to eliminate x_1 from the other two we obtain

$$x_1 - \tfrac{3}{4}x_2 = -\frac{1}{2} \qquad (12')$$

$$-\tfrac{1}{4}x_2 = \frac{2}{5} \qquad (13')$$

$$\tfrac{1}{2}x_2 = a + 1 \qquad (14')$$

Next we use (13') to eliminate x_2 from the other equations, obtaining

$$x_1 + 0 = -8 \qquad (12'')$$

$$x_2 = -10 \qquad (13'')$$

$$0 = a + 6. \qquad (14'')$$

These equations remind us of the situation in Example 3, since we shall be led to a false result unless $a = -6$. We see that equations (12), (13), and (14) have the solution $x_1 = -8$ and $x_2 = -10$ only if $a = -6$. If $a \neq -6$, then there is no solution to these equations.

The examples above illustrate all the possibilities that can occur in the general case. There may be no solutions, exactly one solution, or an infinite number of solutions to a set of simultaneous equations.

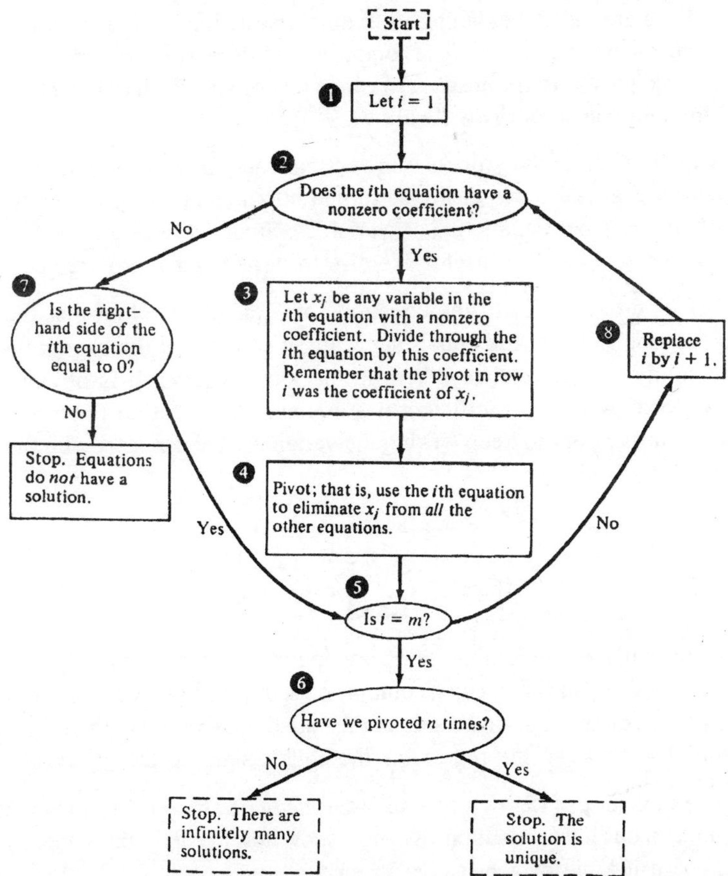

Figure 2.5: Flow diagram for solving m equations in n variables.

The procedure that we have illustrated above is one that turns any set of linear equations into an equivalent set of equations from which the existence of solutions and the solutions can be easily read. A student who learned other ways of solving linear equations may wonder why we use the above procedure—one which is not always the quickest way of solving equations. The answer is that we use it because it always works, that is, it is a *canonical* procedure to apply to any set of linear equations.

The faster methods usually work only for equations that have solutions, and even then may not find all solutions.

The computational process illustrated above is summarised in the flow diagram of Figure 2.5. In that diagram the instructions encircled by dotted lines are either beginning or ending instructions; those enclosed in solid rectangles are intermediate computational steps; and those enclosed in ovals ask questions, the answers to which determine which of two paths the computational process will follow.

The direction of the process is always indicated by arrows. The flow diagram of Figure 2.5 can easily be turned into a computer programme for solving m linear equations in n variables. Students having access to a computer will find it a useful exercise to write such a programme.

Let us return again to the equations of Example 1. Note that the variables, coefficients, and equals signs are in columns at the beginning of the solution and are always kept in the same column. It is obvious that the location of the coefficient is sufficient identification for it and that it is unnecessary to keep writing the variables. We can start with the format or *tableau*

$$\left(\begin{array}{ccc|c} 1 & 4 & 3 & 1 \\ 2 & 5 & 4 & 4 \\ 1 & -3 & -2 & 5 \end{array}\right). \tag{15}$$

Note that the coefficients of x_1 are found in the first column, the coefficients of x_2 in the second column, of x_3 in the third column, and the constants on the right-hand side of the equation all occur in the fourth column. The vertical line represents the equals signs in the equations.

The tableau of (15) will be called the *detached coefficient tableau* for simultaneous linear equations. We now show how to solve simultaneous equations using the detached coefficient tableau.

Example 6: Starting with the tableau of (15) we carry out exactly the same calculations as in Example 1, which lead to the following series of tableaus:

$$\left(\begin{array}{ccc|c} 1 & 4 & 3 & 1 \\ 0 & -3 & -2 & 2 \\ 0 & -7 & -5 & 4 \end{array}\right) \tag{16}$$

$$\left(\begin{array}{ccc|c} 1 & 0 & \frac{1}{3} & \frac{11}{3} \\ 0 & 1 & \frac{2}{3} & -\frac{2}{3} \\ 0 & 0 & -\frac{1}{3} & -\frac{2}{3} \end{array}\right) \tag{17}$$

$$\left(\begin{array}{ccc|c} 1 & 0 & 0 & 3 \\ 0 & 1 & 0 & -2 \\ 0 & 0 & 1 & 2 \end{array}\right). \tag{18}$$

From the tableau of (18) we can easily read the answer $x_1 = 3$, $x_2 = -2$, and $x_3 = 2$, which is the same as before.

The correspondence between the calculations of Example 1 and of the present example is as follows:

(1), (2) and (3) correspond to (15).

(1'), (2') and (3') correspond to (16).

(1"), (2") and (3") correspond to (17).

(1'''), (2''') and (3''') correspond to (18).

Note that in the tableau form we are always careful to keep zero coefficients in each column when necessary.

Example 7: Suppose that we have two sets of simultaneous equations to solve and that they differ only in their right-hand sides. For instance, suppose we want to solve

$$\begin{pmatrix} 1 & 4 & 3 \\ 2 & 5 & 4 \\ 1 & -3 & -2 \end{pmatrix}\begin{pmatrix} x_1 \\ x_2 \\ x_3 \end{pmatrix} = \begin{pmatrix} 1 \\ 4 \\ 5 \end{pmatrix} \text{ and } = \begin{pmatrix} -1 \\ 0 \\ 2 \end{pmatrix}. \tag{19}$$

It is obvious that the calculations on the left-hand side will be the same regardless of the numbers appearing on the right-hand side. Therefore it is possible to solve both sets of simultaneous equations at once. We shall illustrate this in the following series of tableaus:

$$\left(\begin{array}{ccc|cc} 1 & 4 & 3 & 1 & -1 \\ 2 & 5 & 4 & 4 & 0 \\ 1 & -3 & -2 & 5 & 2 \end{array}\right) \tag{20}$$

$$\begin{pmatrix} 1 & 4 & 3 & 1 & -1 \\ 0 & -3 & -2 & 2 & 2 \\ 0 & -7 & -5 & 4 & 3 \end{pmatrix} \tag{21}$$

$$\begin{pmatrix} 1 & 0 & \frac{1}{3} & \frac{11}{3} & \frac{5}{3} \\ 0 & 1 & \frac{2}{3} & -\frac{2}{3} & -\frac{2}{3} \\ 0 & 0 & -\frac{1}{3} & -\frac{2}{3} & \frac{5}{3} \end{pmatrix} \tag{22}$$

$$\begin{pmatrix} 1 & 0 & 0 & 3 & 0 \\ 0 & 1 & 0 & -2 & -4 \\ 0 & 0 & 1 & 2 & 5 \end{pmatrix}. \tag{23}$$

We find the answers

$$x_1 = 3, \qquad x_2 = -2, \qquad x_3 = 2$$

to the first set of equations and the answers

$$x_1 = 0, \qquad x_2 = -4, \qquad x_3 = 5$$

to the second set of equations. The reader should check these answers by substituting into the original equations.

Exercises

1. Find all solutions to the following simultaneous equations:

(a)
$$\begin{aligned} x_1 + 2x_2 + 3x_3 &= 4 \\ 4x_1 + 5x_2 + 6x_3 &= -2 \\ 7x_1 + 8x_2 + 27x_3 &= 10. \end{aligned}$$
[*Ans.* $x_1 = -7,\ x_2 = 4,\ x_3 = 1$.]

(b)
$$\begin{aligned} -x_1 + 3x_3 &= 16 \\ 2x_1 + 7x_2 + 3x_3 &= -6 \\ -x_1 + 7x_2 + 12x_3 &= 41. \end{aligned}$$
[*Ans.* No solution.]

(c)
$$\begin{aligned} -x_1 + 3x_3 &= 8 \\ 2x_1 + 7x_2 + 3x_3 &= -3 \\ -x + 7x_2 + 12x_3 &= 21 \end{aligned}$$

[*Ans.* $x_1 = 3x_3 - 8,\ x_2 = \frac{1}{7}(13 - 9x_3),\ x_3$ = any real number.]

2. Find all solutions of the following simultaneous equations:

(a)
$$\begin{aligned} x_1 + 5x_2 + 3x_3 &= 2 \\ x_1 + x_2 + x_3 &= -2 \\ 13x_1 + 4x_2 + 7x_3 &= 4. \end{aligned}$$

(b)
$$\begin{aligned} 5x_1 + 3x_2 - x_3 &= -2 \\ 2x_1 + 2x_2 + 3x_3 &= 3 \\ 3x_1 + x_3 &= 0. \end{aligned}$$

(c)
$$\begin{aligned} 2x_1 + 3x_2 - x_3 &= 0 \\ -4x_2 + x_3 &= 0 \\ 8x_1 - x_3 &= 0. \end{aligned}$$

3. Rework Examples 2-4 using the detached coefficient tableau.

4. Find all solutions of the following equations using all the detached coefficient tableau:

(a)
$$\begin{aligned} 2x_1 + 2x_2 - x_3 &= -4 \\ -x_2 + 2x_3 &= -2 \\ \frac{x_1}{2} + x_2 + x_3 &= 1. \end{aligned}$$

(b)
$$\begin{aligned} 5x_1 + 3x_2 + 3x_3 &= 7 \\ -2x_1 + 4x_2 + 8x_3 &= -3 \\ -x_1 + 3x_2 + 4x_3 &= 2. \end{aligned}$$

(c)
$$\begin{aligned} 3x_1 + 6x_2 - x_3 &= -7 \\ x_1 + 5x_2 &= 0 \\ 2x_2 + x_3 &= 14. \end{aligned}$$

5. Find all solutions of the following equations:

(a)
$$\begin{aligned} -x_2 + x2 - x_3 + x_4 &- 13 \\ 2x_2 - 6x_3 + 7x_4 &= 0 \\ -2x_1 + x_2 + x_3 + 5x_4 &= -13 \\ 3x_1 - 2x_2 + 2x_3 + x_4 &= 26. \end{aligned}$$

(b)
$$\begin{aligned} x_1 + 2x_2 + 3x_3 + 4x_4 &= 10 \\ 2x_1 - x_2 + x_3 - x_4 &= 1 \\ 3x_1 + x_2 + 4x_3 + 3x_4 &= 11 \\ -2x_1 + 6x_2 + 4x_3 + 10x_4 &= 18 \end{aligned}$$

6. Solve the following four simultaneous sets whose right-hand sides are listed under (a), (b), (c), and (d) below. Use the detached coefficient tableau.

	(a)	(b)	(c)	(d)
$3x_1 + 6x_2 + 4x_3 =$	–4	2	1	–1
$4x_1 + 8x_2 + 5x_3 =$	8	2	1	–1
$4x_1 + 3x_3 =$	0	2	1	1.

[*Ans.* (a) $x_1 = 30$, $x_2 = 11$, $x_3 = -40$.]

7. Solve the following four sets of simultaneous equations, which differ only in their right-hand sides:

	(a)	(b)	(c)	(d)
$-x_1 + 2x_3 =$	–2	–2	4	1
$2x_1 + 7x_2 + 3x_3 =$	4	2	–1	12
$-x_1 + 7x_2 + 11x_3 =$	0	4	1	17.

[*Ans.* (d) $x_1 = x_2 = x_3 = 1$.]

8. Solve the following four sets of simultaneous equations:

	(a)	(b)	(c)	(d)
$x_1 + 6x_2 + x_3 =$	12	12	0	–9
$-x_1 + 2x_2 - 9x_3 =$	–9	0	12	0
$3x_2 + 3x_3 =$	0	–9	–9	12.

[*Ans.*(a) $x_1 = 10\frac{3}{4}$, $x_2 = \frac{1}{4}$, $x_3 = -\frac{1}{4}$.]

9. A man is ordered by his doctor to take 10 units of vitamin A, 9 units of vitamin D, and 19 units of vitamin E each day. The man can choose from three brands of vitamin pills. Brand X contains two units of vitamin A, three units of vitamin D, and five units of

vitamin E; brand Y has 1, 3, and 4 units, respectively; and brand Z has 1 unit of vitamin A, 1 of vitamin E, and none of vitamin D.

(a) Find all possible combinations of pills that will provide exactly the required amounts of vitamins.

(b) If brand X costs ¢1 a pill, brand Y ¢6, and brand Z ¢3, is there a solution costing exactly ¢15 a day?

(c) What is the least expensive solution? The most expensive?

10. Show that the equations

$$4x_1 - 4x_2 + ax_3 = c$$
$$3x_1 - 2x_2 + bx_3 = d$$

always have a solution for all values of *a, b, c,* and *d.*

11. Find conditions on *a, b,* and *c* in order that the equations

$$3x_1 + 2x_2 = a$$
$$-4x_1 + x_2 = b$$
$$2x_1 + 5x_2 = c$$

have a solution.

12. For what value of the constant *k* does the following system have a unique solution? Find the solution in this case. What is the case if *k* does not take on this value?

$$2x + 4z = 6$$
$$3x + y + z = -1$$
$$2y - z = -2$$
$$x - y + kz = -5.$$

[*Ans.* $k = -2$; $x = -1$, $y = 0$, $z = 2$; no solution.]

13. Let $x = (x_1, x_2, x_3)$, let $A = \begin{pmatrix} a \\ b \\ c \end{pmatrix}$, and let d be any number. What can you say about the truth set of the statement $xA = d$ in the following cases:

(a) $A \neq 0$?

(b) $A = 0$, $d = 0$?

(c) $A = 0,\ d \neq 0$?

What can you conclude?

14. Let P be the matrix $P = \begin{pmatrix} \frac{1}{2} & \frac{1}{2} \\ \frac{1}{4} & \frac{3}{2} \end{pmatrix}$, and let $x = (x_1, x_2)$.

(a) Find all solutions of the equation $xP = x$.

(b) Choose the solution(*s*) for which $x_1 + x_2 = 1$.

15. Let P be the matrix $P = \begin{pmatrix} \frac{1}{2} & \frac{1}{2} & 0 \\ \frac{1}{3} & \frac{1}{3} & \frac{1}{3} \\ \frac{1}{4} & \frac{1}{2} & \frac{1}{4} \end{pmatrix}$, and let $x = (x_1\ x_2,\ x_3)$.

(a) Find all solutions of the equation $xP - x$.

(b) Choose the solution(*s*) for which $x_1 + x_2 + x_3 = 1$.

16. Let P be as in Exercise 15, and let $x = \begin{pmatrix} x_1 \\ x_2 \\ x_3 \end{pmatrix}$. Redo the exercise, using the equation $Px = x$.

17. Let x be as in Exercise 15, and let A be the matrix

$$\begin{pmatrix} 2 & 0 & 3 \\ 0 & 1 & 5 \\ -3 & 0 & 1 \end{pmatrix}.$$

(a) Find all solutions of the equation $xA = x$.

(b) Choose the solution(*s*) for which $x_1 + x_2 + x_3 = 1$.

[*Ans.* $x_1 = \frac{15}{11}, x_2 = -\frac{9}{11}, x_3 = \frac{5}{11}$.]

18. (a) Show that the simultaneous linear equations

$$3x_1 - 5x_2 + 3x_3 = 9$$
$$4x_1 + 4x_2 - 7x_3 = 0$$

can be interpreted as a single-matrix-times-column-vector equation of the form

$$\begin{pmatrix} 3 & -5 & 3 \\ 4 & 4 & -7 \end{pmatrix} \begin{pmatrix} x_1 \\ x_2 \\ x_3 \end{pmatrix} = \begin{pmatrix} 9 \\ 0 \end{pmatrix}.$$

(b) Show that any set of simultaneous Linear Equations may be interpreted as a matrix equation of the form $Ax = b$, where A is an $m \times n$ matrix, x is an n-component column vector, and b is an m-component column vector.

19. (a) Show that the equations of Exercise 18(a) can be interpreted as a row-vector-times-matrix equation of the form

$$(x_1, x_2, x_3)\begin{pmatrix} 3 & 4 \\ -5 & 4 \\ 3 & -7 \end{pmatrix} = (9, 0).$$

(b) Show that any set of simultaneous Linear Equations may be interpreted as a matrix equation of the form $xA = b$, where A is an $m \times n$ matrix, x is an m-component row vector, and b is an n-component row vector.

20. (a) Show that the simultaneous Linear Equations of Exercise 18(a) can be interpreted as asking for all possible ways of expressing the column vector $\begin{pmatrix} 9 \\ 0 \end{pmatrix}$ in terms of the column vectors $\begin{pmatrix} 3 \\ 4 \end{pmatrix}$, $\begin{pmatrix} -5 \\ 4 \end{pmatrix}$ and $\begin{pmatrix} 3 \\ -7 \end{pmatrix}$.

(b) Show that any set of Linear Equations may be interpreted as asking for all possible ways of expressing a column vector in terms of given column vectors.

21. Redo Exercise 20, using Exercise 19(a) and *Row* Vectors.

22. Consider the following set of simultaneous equations:

$$x_1 - x_2 = a$$
$$x_3 + x_4 = b$$
$$x_2 - x_4 = c$$
$$x_1 + x_3 = d.$$

(a) For what conditions on a, b, c, and d will these equations have a solution?

(b) Give a set of values for a, b, c, and d for which the equations do not have a solution.

(c) Show that if there is one solution to these equations, then there are infinitely many solutions.

23. Which of the following statements are true and which false concerning the solution of m simultaneous Linear Equations in n unknowns written in the form $Ax = b$?

    (a) If there are infinitely many solutions, then $n > m$.

    (b) If the solution is unique, then $n = m$.

    (c) If $m = n$, then the solution is unique.

    (d) If $n > m$, then there cannot be a unique solution.

    (e) If $b = 0$, then there is always at least one solution.

    (f) If $b = 0$, then there are always infinitely many solutions.

    (g) If $b = 0$, and $x^{(1)}$ and $x^{(2)}$ are solutions, then $x^{(1)} + x^{(2)}$ is also a solution.

[*Ans.* (d), (e), and (g) are true.]

Inverse Square Matrix

If A is a square matrix and B is another square matrix of the same size having the property that $BA = I$ (where I is the identity matrix), then we say that B is the *inverse* of A. When it exists, we shall denote the inverse of A by the symbol A^{-1}. To give a numerical example, let A and A^{-1} be the following:

$$A = \begin{pmatrix} 4 & 0 & 5 \\ 0 & 1 & -6 \\ 3 & 0 & 4 \end{pmatrix} \quad (1)$$

$$A^{-1} = \begin{pmatrix} 4 & 0 & -5 \\ -18 & 1 & 24 \\ -3 & 0 & 4 \end{pmatrix}. \quad (2)$$

Then we have

$$A^{-1}A = \begin{pmatrix} 4 & 0 & -5 \\ -18 & 1 & 24 \\ -3 & 0 & 4 \end{pmatrix} \cdot \begin{pmatrix} 4 & 0 & 5 \\ 0 & 1 & -6 \\ 3 & 0 & 4 \end{pmatrix} = \begin{pmatrix} 1 & 0 & 0 \\ 0 & 1 & 0 \\ 0 & 0 & 1 \end{pmatrix} = I$$

If we multiply these matrices in the other order, we also get the identity matrix; thus

$$AA^{-1} = \begin{pmatrix} 4 & 0 & 5 \\ 0 & 1 & -6 \\ 3 & 0 & 4 \end{pmatrix} \cdot \begin{pmatrix} 4 & 0 & -5 \\ -18 & 1 & 24 \\ -3 & 0 & 4 \end{pmatrix} = \begin{pmatrix} 1 & 0 & 0 \\ 0 & 1 & 0 \\ 0 & 0 & 1 \end{pmatrix} = I.$$

In general it can be shown that if A is a square matrix with inverse A^{-1}, then the inverse satisfies the equation

$$A^{-1}A = AA^{-1} = I.$$

Next we show that a square matrix can have only one inverse. For suppose that in addition to A^{-1} we also have a B such that

$$BA = I.$$

Then we see that

$$B = BI = B(AA^{-1}) = (BA)A^{-1} = IA^{-1} = A^{-1}.$$

Finding the inverse of a matrix is analogous to finding the reciprocal of an ordinary number, but the analogy is not complete. Every non-zero number has a reciprocal, but there are matrices, not the zero matrix, which have no inverse. For example, if

$$A = \begin{pmatrix} 1 & -1 \\ -1 & 1 \end{pmatrix} \text{ and } B = \begin{pmatrix} 1 & 1 \\ 1 & 1 \end{pmatrix},$$

then

$$AB = \begin{pmatrix} 1 & -1 \\ -1 & 1 \end{pmatrix} \cdot \begin{pmatrix} 1 & 1 \\ 1 & 1 \end{pmatrix} = \begin{pmatrix} 0 & 0 \\ 0 & 0 \end{pmatrix} = 0.$$

From this we shall show that neither A nor B can have an inverse. To show that A does not have an inverse, let us assume that A had an inverse A^{-1}. Then

$$B = (A^{-1}A)B = A^{-1}(AB) = A^{-1}0 = 0,$$

contradicting the fact that $B \neq 0$. The proof that B cannot have an inverse is similar.

Let us now try to calculate the inverse of the matrix A in (1). Specifically, let's try to calculate the first column of A^{-1}. Let

$$x = \begin{pmatrix} x_1 \\ x_2 \\ x_3 \end{pmatrix}$$

be the desired entries of the first column. Then from the equation $AA^{-1} = I$ we see that we must solve

$$\begin{pmatrix} 4 & 0 & 5 \\ 0 & 1 & -6 \\ 3 & 0 & 4 \end{pmatrix} \begin{pmatrix} x_1 \\ x_2 \\ x_3 \end{pmatrix} = \begin{pmatrix} 1 \\ 0 \\ 0 \end{pmatrix}.$$

Similarly, to find the second and third columns of A^{-1} we want to solve the additional sets of equations,

$$\begin{pmatrix} 4 & 0 & 5 \\ 0 & 1 & -6 \\ 3 & 0 & 4 \end{pmatrix} \begin{pmatrix} x_1 \\ x_2 \\ x_3 \end{pmatrix} = \begin{pmatrix} 0 \\ 1 \\ 0 \end{pmatrix} \text{ and } = \begin{pmatrix} 0 \\ 0 \\ 1 \end{pmatrix},$$

respectively. We thus have three sets of simultaneous equations that differ only in their righ-thand sides.

To solve them, we start with the tableau

$$\left(\begin{array}{ccc|ccc} 4 & 0 & 5 & 1 & 0 & 0 \\ 0 & 1 & -6 & 0 & 1 & 0 \\ 3 & 0 & 4 & 0 & 0 & 1 \end{array}\right) \tag{3}$$

This gives rise to the following series of tableaus. In (3) divide the first row by 4, copy the second row, and subtract 3 times the new first row from the old third row, which yields the tableau

$$\left(\begin{array}{ccc|ccc} 1 & 0 & \frac{5}{4} & \frac{1}{4} & 0 & 0 \\ 0 & 1 & -6 & 0 & 1 & 0 \\ 0 & 0 & \frac{1}{4} & -\frac{3}{4} & 0 & 1 \end{array}\right). \tag{4}$$

Next we multiply the third row of (4) by 4, multiply the new third row by 6 and add to the old second row, and multiply the new third row by $\frac{5}{4}$ and subtract from the old first row. We have the final tableau:

$$\left(\begin{array}{ccc|ccc} 1 & 0 & 0 & 4 & 0 & -5 \\ 0 & 1 & 0 & -18 & 1 & 24 \\ 0 & 0 & 1 & -3 & 0 & 4 \end{array}\right). \tag{5}$$

We see that the inverse A^{-1} which is given in (2) appears to the right of the vertical line in the tableau of (5).

The procedure just illustrated will find the inverse of any square matrix A, providing A has an inverse. We summarise it as follows:

Rule for Inverting a Matrix: Let A be a matrix that has an inverse. To find the inverse of A start with the tableau

$$(A \backslash I)$$

and change it by row transformations into the tableau

$$(I \backslash B).$$

The resulting matrix B is the inverse A^{-1} *of* A.

Even if A has no inverse, the procedure just outlined can be started. At some point in the procedure a tableau will be found that is not of the desired final form and from which it is impossible to change by row transformations of the kind described.

Example 1: Show that the matrix

$$A = \begin{pmatrix} 4 & 0 & 8 \\ 0 & 1 & -6 \\ 2 & 0 & 4 \end{pmatrix}$$

has no inverse.

We set up the initial tableau as follows:

$$\left(\begin{array}{ccc|ccc} 4 & 0 & 8 & 1 & 0 & 0 \\ 0 & 1 & -6 & 0 & 1 & 0 \\ 2 & 0 & 4 & 0 & 0 & 1 \end{array}\right). \tag{6}$$

Carrying out one set of row transformations, we obtain the second tableau as follows:

$$\left(\begin{array}{ccc|ccc} 1 & 0 & 2 & \frac{1}{4} & 0 & 0 \\ 0 & 1 & -6 & 0 & 1 & 0 \\ 0 & 0 & 0 & -\frac{1}{2} & 0 & 1 \end{array}\right). \tag{7}$$

It is clear that we cannot proceed further since there is a row of zeros to the left of the equals sign on the third set of equations. Hence, we conclude that A has no inverse.

Because of the form of the final tableau in (7), we see that it is impossible to solve the equations

$$\begin{pmatrix} 4 & 0 & 8 \\ 0 & 1 & -6 \\ 2 & 0 & 4 \end{pmatrix} \begin{pmatrix} x_1 \\ x_2 \\ x_3 \end{pmatrix} = \begin{pmatrix} 0 \\ 0 \\ 1 \end{pmatrix},$$

since these equations are inconsistent as is shown by the tests developed in 'Solution of Linear Equations'. In other words, it is not possible to solve for the third column of the inverse matrix.

It is clear that an $n \times n$ matrix A has an inverse if and only if the following sets of simultaneous equations

$$Ax = \begin{pmatrix} 1 \\ 0 \\ \vdots \\ 0 \end{pmatrix}, \quad Ax = \begin{pmatrix} 0 \\ 1 \\ \vdots \\ 0 \end{pmatrix}, \quad \ldots, \quad Ax = \begin{pmatrix} 0 \\ 0 \\ \vdots \\ 1 \end{pmatrix}$$

can all be uniquely solved. And these sets of simultaneous equations, since they all share the same left-hand sides, can be solved uniquely if and only if the transformation of the rule for inverting a matrix can be carried out. Hence, we have proved the following theorem.

Theorem: A square matrix A has an inverse if and only if the tableau

$$(A \backslash I)$$

can be transformed by row transformations into the tableau

$$(I \backslash A^{-1}).$$

Example 2: Let us find the inverse of the matrix

$$A = \begin{pmatrix} 1 & 4 & 3 \\ 2 & 5 & 4 \\ 1 & -3 & -2 \end{pmatrix}.$$

The initial tableau is

$$\left(\begin{array}{ccc|ccc} 1 & 4 & 3 & 1 & 0 & 0 \\ 2 & 5 & 4 & 0 & 1 & 0 \\ 1 & -3 & -2 & 0 & 0 & 1 \end{array}\right).$$

Transforming it by row transformations, we obtain the following series of tableaus:

$$\left(\begin{array}{ccc|ccc} 1 & 4 & 3 & 1 & 0 & 0 \\ 0 & -3 & -2 & -2 & 1 & 0 \\ 0 & -7 & -5 & -1 & 0 & 1 \end{array}\right)$$

$$\left(\begin{array}{ccc|ccc} 1 & 0 & \frac{1}{3} & -\frac{5}{3} & \frac{4}{3} & 0 \\ 0 & 1 & \frac{2}{3} & \frac{2}{3} & -\frac{1}{3} & 0 \\ 1 & 0 & -\frac{1}{3} & \frac{11}{3} & -\frac{7}{3} & 1 \end{array}\right)$$

$$\left(\begin{array}{ccc|ccc} 1 & 0 & 0 & 2 & -1 & 1 \\ 0 & 1 & 0 & 8 & -5 & 2 \\ 0 & 0 & 1 & -11 & 7 & -3 \end{array}\right).$$

The inverse of A is then

$$A^{-1} = \begin{pmatrix} 2 & -1 & 1 \\ 8 & -5 & 2 \\ -11 & 7 & -3 \end{pmatrix}.$$

The reader should check that $A^{-1}A = AA^{-1} = I$.

Example 3: *A* cookie recipe requires 4 cups of sugar and 2 cups of flour while a cake recipe needs 3 cups of sugar and 4 cups of flour. If we have 40 cups of sugar and 30 cups of flour on hand, how many recipes of each can we make? In order to answer this question let x be the number of cookie recipes and y the number of cake recipes to be made. Then the sugar and flour requirements give rise to the following two equations:

$$4x + 3y = 40 \text{ (Sugar equation)}$$

$$2x + 4y = 30 \text{ (Flour equation).}$$

Let us rewrite these equations in matrix form as

$$\begin{pmatrix} 4 & 3 \\ 2 & 4 \end{pmatrix}\begin{pmatrix} x \\ y \end{pmatrix} = \begin{pmatrix} 40 \\ 30 \end{pmatrix}.$$

If we can invert the matrix we can solve the problem as

$$\begin{pmatrix} x \\ y \end{pmatrix} = \begin{pmatrix} 4 & 3 \\ 2 & 4 \end{pmatrix}^{-1}\begin{pmatrix} 40 \\ 30 \end{pmatrix}.$$

The initial tableau of the matrix inversion problem is

$$\left(\begin{array}{cc|cc} 4 & 3 & 1 & 0 \\ 2 & 4 & 0 & 1 \end{array}\right).$$

Pivoting on the 4 in the upper left-hand corner gives

$$\left(\begin{array}{cc|cc} 1 & \frac{3}{4} & \frac{1}{4} & 0 \\ 0 & \frac{5}{2} & -\frac{1}{2} & 1 \end{array}\right).$$

Finally, pivoting on the $\frac{5}{2}$ term we obtain

$$\left(\begin{array}{cc|cc} 1 & 0 & \frac{2}{5} & -\frac{3}{10} \\ 0 & 1 & -\frac{1}{5} & \frac{2}{5} \end{array}\right),$$

and so the inverse of the original matrix appears on the right. The solution to our problem is, then,

$$\begin{pmatrix} x \\ y \end{pmatrix} = \begin{pmatrix} \frac{2}{5} & -\frac{3}{10} \\ -\frac{1}{5} & \frac{2}{5} \end{pmatrix} \begin{pmatrix} 40 \\ 30 \end{pmatrix} = \begin{pmatrix} 7 \\ 4 \end{pmatrix}.$$

In other words, we can make 7 batches of cookies and 4 cakes from the materials we have.

Exercises

1. Compute the inverse of each of the following matrices:

$$A = \begin{pmatrix} 3 & -1 \\ -1 & 2 \end{pmatrix}, \quad B = \begin{pmatrix} -5 & 1 & 10 \\ 9 & -2 & -17 \\ -4 & 1 & 8 \end{pmatrix},$$

$$C = \begin{pmatrix} 3 & 1 & 0 \\ 2 & 2 & 2 \\ 11 & 3 & 1 \end{pmatrix}, \quad D = \begin{pmatrix} 1 & -7 & 3 & 2 \\ 0 & -1 & 7 & 3 \\ 0 & 0 & -1 & 2 \\ 0 & 0 & 0 & 1 \end{pmatrix}.$$

[*Partial Ans:* $A^{-1} = \begin{pmatrix} .4 & .2 \\ .2 & .6 \end{pmatrix}, \quad B^{-1} = \begin{pmatrix} 1 & 2 & 3 \\ -4 & 0 & 5 \\ 1 & 1 & 1 \end{pmatrix}$.]

2. Let B and C be the matrices of Exercise 1; let $x = \begin{pmatrix} x_1 \\ x_2 \\ x_3 \end{pmatrix}$ and $y =$ (y_1, y_2, y_3); let b, c, d, and e be the following vectors:

$$a = \begin{pmatrix} -1 \\ 0 \\ 2 \end{pmatrix},\ b = \begin{pmatrix} 4 \\ -1 \\ 3 \end{pmatrix},\ c = (1, 5, 3),\ \mathrm{d} = (1,1,1),\ \text{and } e = \begin{pmatrix} 1 \\ 1 \\ 1 \end{pmatrix}.$$

Uses the inverses computed in Exercise 1 to solve the following equations:

(a) $Bx = a.$

(b) $yB = d.$

(c) $Cx = e.$

(d) $Bx = b.$

(e) $yB = c.$

(f) $yC = c.$

3. Show that each of the following matrices fails to have an inverse.

$$A = \begin{pmatrix} 4 & 2 \\ 6 & 3 \end{pmatrix},\ B = \begin{pmatrix} 3 & 1 & 0 \\ 8 & 2 & 1 \\ 11 & 3 & 1 \end{pmatrix},\ C = \begin{pmatrix} -3 & -1 & 1 \\ 4 & 3 & -2 \\ -7 & 1 & 1 \end{pmatrix},$$

$$D = \begin{pmatrix} 2 & 1 & 1 & 1 \\ 1 & -3 & 5 & -2 \\ 0 & 3 & -4 & 2 \\ 2 & 0 & 1 & -1 \end{pmatrix}.$$

4. For each of the matrices in Exercise 3 find a non-zero vector whose product with the given matrix is 0.

5. Solve the following four sets of simultaneous equations by first writing them in the form $Ax = B$, where B is a 3×4 matrix, and finding the inverse of A.

	(a)	(b)	(c)	(d)
$4x_1 + 5x_3 =$	1	1	0	8

$$x_2 - 6x_3 = \quad 2 \quad 0 \quad 0 \quad 1$$

$$3x_1 + 4x_3 = \quad 3 \quad 0 \quad 1 \quad 0.$$

[*Ans.* (a) $x_1 = -11$, $x_2 = 56$, $x_3 = 9$.]

6. Let A be a square matrix. Show that if A has no inverse, then neither do any of its positive powers A^k. Show that if A has an inverse, then the inverse of A^2 is $(A^{-1})^2$. What is the inverse of A^3? Of A^n?
7. The formula $(A^{-1})^{-1} = A$ states that if A^{-1}, has an inverse A^{-1} then A^{-1} itself has an inverse and this inverse is A. Prove both parts of this statement.
8. Expand the formula $(AB)^{-1} = B^{-1}A^{-1}$ into a two-part statement analogous to the one in Exercise 7. Then prove both parts of your statement.
9. Give a criterion for deciding whether the 2×2 matrix $\begin{pmatrix} a & b \\ c & d \end{pmatrix}$ has an inverse.

[*Ans.* $ad \neq bc$.]

10. Give a formula for $\begin{pmatrix} a & b \\ c & d \end{pmatrix}^{-1}$, when it exists.
11. If $\begin{pmatrix} a & b \\ c & d \end{pmatrix}$ has an inverse and has integer components, what condition must it fulfil in order that $\begin{pmatrix} a & b \\ c & d \end{pmatrix}^{-1}$ have integer components?
12. Let A be the matrix $\begin{pmatrix} 5 & 3 \\ 3 & 2 \end{pmatrix}$.

    (a) Use Exercise 10 to find A^{-1}.

    (b) Use the result of (a) to solve the equations $Ax = b$ and $A^2x = c$, where $x = \begin{pmatrix} x_1 \\ x_2 \end{pmatrix}$, $b = \begin{pmatrix} -1 \\ 0 \end{pmatrix}$, and $c = \begin{pmatrix} 1 \\ 1 \end{pmatrix}$.
13. (a) Show that $(AB)^{-1} \neq A^{-1}B^{-1}$ for the matrices $A = \begin{pmatrix} 2 & -5 \\ -1 & 3 \end{pmatrix}$ and

$$B = \begin{pmatrix} -1 & 0 \\ 2 & 1 \end{pmatrix}.$$

(b) Find $(AB)^{-1}$ in two ways. [*Hint:* Use Exercises 10 and 8.]

14. Solve the following problems by first inverting the matrix involved.

 (a) An automobile factory produces two models. The first requires 1 man-hour to paint and $\frac{1}{2}$ man-hour to polish; the second requires 1 man-hour for each process. During each hour that the assembly line is operating, there are 100 man-hours available for painting and 80 man-hours for polishing. How many of each model can be produced each hour if all the man-hours available are to be utilised?

 (b) Suppose each car of the first type requires 10 widgets and 14 shims, and each car of the second type requires 7 widgets and 10 shims. The factory can obtain 800 widgets and 1,130 shims each hour. How many cars of each model can it produce while utilising all the parts available?

 [*Ans.* 45, 50.]

15. Solve the following problem by first inverting the matrix. (Assume $ad \neq bc$.) If a grinding machine is supplied x pounds of meat and y pounds of scraps (meat scraps and fat) per day, then it will produce $ax + by$ pounds of ground meat and $cx + dy$ pounds of hamburger per day. In other words, its production vector is

$$\begin{pmatrix} a & b \\ c & d \end{pmatrix}\begin{pmatrix} x \\ y \end{pmatrix}.$$

 What inputs are necessary in order to get 25 pounds of ground meat and 70 pounds of hamburger? In order to get 20 pounds of ground meat and 100 pounds of hamburger?

16. A square matrix is *lower-triangular* if it has zeros on and above its main diagonal. For instance, $Q = \begin{pmatrix} 0 & 0 & 0 \\ -1 & 0 & 0 \\ 4 & 3 & 0 \end{pmatrix}$ is lower-triangular.

 (a) Compute Q^2.

 (b) Compute Q^3.

 (c) Show that $Q^k = 0$ for $k \geq 3$.

17. Let Q be as in Exercise 16.

    (a) Show that $(I - Q)(I + Q + Q^2) = I - Q^3 = I$.

    (b) Show that, because of (a), $I + Q + Q^2 = (I - Q)^{-1}$.

    (c) Use (b) to compute $(I - Q)^{-1}$.

    (d) Let $w = (w_1, w_2, w_3)$, $d = (-1, 5, 3)$. Use (c) to solve the equation $w = wQ + d$.

18. (a) Show that the sum of any two lower-triangular matrices is lower-triangular.

    (b) Show that the product of any two lower-triangular matrices is lower-triangular.

19. Let Q be an $n \times n$ lower-triangular matrix.

    (a) Show that $Q^k = 0$ for $k \geq n$.

    (b) Show that $(I - Q)(I + Q + \cdots + Q^{n-1}) = I - Q^n = I$.

    (c) Show that $(I - Q)^{-1} = I + Q + \cdots + Q^{n-1}$.

    (d) Show that all entries above the main diagonal of $(I - Q)^{-1}$ are 0.

    (e) Show that if Q has non-negative *integer* entries, then so does $(I - Q)^{-1}$.

20. Find $(I - Q)^{-1}$ for each of the following:

    (a) $\begin{pmatrix} 0 & 0 \\ 0 & 0 \end{pmatrix}$.

    (b) $\begin{pmatrix} 0 & 0 & 0 \\ 3 & 0 & 0 \\ -1 & 2 & 0 \end{pmatrix}$.

    (c) $\begin{pmatrix} 0 & 0 & 0 & 0 \\ 5 & 0 & 0 & 0 \\ 4 & -1 & 0 & 0 \\ -2 & 1 & 3 & 0 \end{pmatrix}$.

(d) $\begin{pmatrix} 0 & 0 & 0 & 0 & 0 \\ 3 & 0 & 0 & 0 & 0 \\ 0 & -1 & 0 & 0 & 0 \\ 1 & 5 & 4 & 0 & 0 \\ -2 & 1 & 2 & -3 & 0 \end{pmatrix}$.

Applications of Matrix Theory

In this section we shall show applications of matrix theory to Markov chains. For simplicity we shall confine our discussion to three-state Markov chains, but a similar procedure will work for any other Markov chain.

We noted that to each Markov chain there was a matrix of transition probabilities. For example, if there are three states, a_1, a_2, and a_3, then

$$P = \begin{array}{c} \\ a_1 \\ a_2 \\ a_3 \end{array} \begin{array}{c} \begin{array}{ccc} a_1 & a_2 & a_3 \end{array} \\ \begin{pmatrix} p_{11} & p_{12} & p_{13} \\ p_{21} & p_{22} & p_{23} \\ p_{31} & p_{32} & p_{33} \end{pmatrix} \end{array}$$

Is the transition matrix for the chain. Recall that the *row sums* of P are all equal to 1. Such a matrix is called a *transition matrix*.

Definition: A *transition matrix* is a square matrix with non-negative entries such that the sum of the entries in each row is 1.

In order to obtain a Markov chain we must specify how the process starts. Suppose that the initial state is chosen by a chance device that selects state a_j with probability $p_j^{(0)}$. We can represent these initial probabilities by means of the vector $p^{(0)} = \left(p_1^{(0)}, p_2^{(0)}, p_3^{(0)}\right)$. We can construct a tree measure for as many steps of the process as we wish to consider. Let $p_j^{(n)}$ be the probability that the process will be in state a_j after n steps. Let the vector of these probabilities be $p^{(n)} = \left(p_1^{(n)}, p_2^{(n)}, p_3^{(n)}\right)$.

Definition: A row vector is called a *probability vector* if it has non-negative components whose sum is 1.

Obviously the vectors $p^{(0)}$ and $p^{(n)}$ are probability vectors. Also each row of a transition matrix is a probability vector.

By means of the tree measure it can be shown that these probabilities satisfy the following equations:

$$p_1^{(n)} = p_1^{(n-1)} p_{11} + p_2^{(n-1)} p_{21} + p_3^{(n-1)} p_{31},$$

$$p_2^{(n)} = p_1^{(n-1)} p_{12} + p_2^{(n-1)} p_{22} + p_3^{(n-1)} p_{32},$$

$$p_3^{(n)} = p_1^{(n-1)} p_{13} + p_2^{(n-1)} p_{23} + p_3^{(n-1)} p_{33}.$$

It is not hard to give intuitive meanings to these equations. The first one, for example, expresses the fact that the probability of being in state a_l after n steps is the sum of the probabilities of being at each of the three possible states after $n - 1$ steps and then moving to state a_1 on the nth step. The interpretation of the other equations is similar.

If we recall the definition of the product of a vector times a matrix, we can write the equations above as

$$p^{(n)} = p^{(n-1)}p.$$

If we substitute values of n, we get the equations: $p^{(1)} = p^{(0)}P;p^{(2)} - p^{(1)}P = p^{(0)}P^2$; $p^{(3)} = p^{(2)}P - p^{(0)}P^3$; and so on. In general, it is evident that

$$p^{(n)} = p^{(0)pn}$$

Thus, we see that, if we multiply the vector $p^{(0)}$ of initial probabilities by the nth power of the transition matrix P, we obtain the vector $p^{(n)}$, whose components give the probabilities of being in each of the states after n steps.

In particular, let us choose $p^{(0)} = (1, 0, 0)$, which is equivalent to letting the process start in state a_1. From the equation above we see that then $p^{(n)}$ is the first row of the matrix p^n. Thus, the elements of the first row of the matrix P^n give us the probabilities that after n steps the process will be in a given one of the states, under the assumption that it started in state a_1. In the same way, if we choose $p^{(0)} = (0, 1, 0)$, we see that the second row of P^n gives the probabilities that the process will be in one of the various states after n steps, given that it started in state a_2. Similarly the third row gives these probabilities, assuming that the process started in state a_3.

We considered special Markov chains that started in given fixed states. There we arrived at a matrix $p^{(n)}$ whose ith row gave the probabilities of the process ending in the various states, given that it started at state a_i. By comparing the work that we did there with what we have just done, we see that the matrix $p^{(n)}$ is merely the nth power of P, that is, $P^{(n)} = P^n$. Matrix multiplication thus gives a convenient way of computing the desired probabilities.

Definition: The probability vector w is a *fixed point* of the matrix P, if $w = wP$.

Example 1: Consider the transition matrix

$$P = \begin{pmatrix} \frac{2}{3} & \frac{1}{3} \\ \frac{1}{2} & \frac{1}{2} \end{pmatrix} = \begin{pmatrix} .667 & .333 \\ .500 & .500 \end{pmatrix}.$$

If $w = (.6, .4)$, then we see that

$$wP = (.6, .4)\begin{pmatrix} \frac{2}{3} & \frac{1}{3} \\ \frac{1}{2} & \frac{1}{2} \end{pmatrix} = (.6, .4) = w,$$

so that w is the fixed point of the matrix P.

If we had happened to choose the vector w as our initial probability vector $p^{(0)}$, we would have had $p^{(n)} = p^{(0)}p^n = wP^n = w = p^{(0)}$. In this case the probability of being at any particular state is the same at all steps of the process. Such a process is in *equilibrium.*

As seen above, in the study of Markov chains we are interested in the powers of the matrix P. To see what happens to these powers, let us further consider the example.

Example 1 continued: Suppose that we compute powers of the matrix P in the example above. We have

$$p^2 = \begin{pmatrix} .611 & .389 \\ .583 & .417 \end{pmatrix}, \quad p^3 = \begin{pmatrix} .602 & .398 \\ .597 & .403 \end{pmatrix}, \text{ and so on.}$$

It looks as if the matrix P^n is approaching the matrix

$$W = \begin{pmatrix} .6 & .4 \\ .6 & .4 \end{pmatrix};$$

and, in fact, it can be shown that this is the case. (When we say that P^n approaches W we mean that each entry in the matrix P^n gets close to the corresponding entry in W.) Note that each row of W is the fixed point w of the matrix P.

Definition: A transition matrix is said to be *regular* if some power of the matrix has only positive components.

Thus, the matrix in the example is regular, since every entry in it is positive, so that the first power of the matrix has all positive entries. Other examples occur in the exercises.

Theorem: If P is a regular transition matrix, then:

(i) the powers P^n approach a matrix W;

(ii) each row of W is the same probability vector w;

(iii) the components of w are positive.

We omit the proof of this theorem; however, we can prove the next theorem.

Theorem: If P is a regular transition matrix, and W and w are given by the previous theorem, then:

(a). if p is any probability vector, pP^n approaches w;

(b). the vector w is the unique fixed-point probability vector of P.

Proof: First let us consider the vector pW. The first column of W has a w_1 in each row. Hence, in the first component of pW each component of p is multiplied by w_1 and therefore we have w_1 times the sum of the components of p_1, which is w_1. Doing the same for the other components, we note that pW is simply w. But pP^n approaches pW; hence it approaches w. Thus, if any probability vector is multiplied repeatedly by P, it approaches the fixed point w. This proves part (a).

Since the powers of P approach W, $P^{n+1} = P^nP$ approaches W, but it also approaches WP; hence $WP = W$. Any one row of this matrix equation states that $wP = w$; hence w is a fixed point (and by the previous theorem a probability vector). We must still show that it is unique. Let u be any probability-vector fixed point of P. By part a of the theorem we know that uP^n approaches w. But since u is (a) fixed point, $uP^n = u$. Hence u remains fixed but "approaches" w. This is possible only if $u = w$. Hence, w is the only probability-vector fixed point. This completes the proof of part (b).

The following is an important consequence of the above theorem. If we take as p the vector $p^{(0)}$ of initial probabilities, then the vector $pP^n = p^{(n)}$ gives the probabilities after n steps, and this vector approaches w. Therefore, no matter what the initial probabilities are, if P is regular, then after a large number of steps the probability that the process is in state a_j will be very nearly w_j.

We noted for an independent trials process that if p is the probability of a given outcome a, then this may be given an alternate interpretation by means of the law of large numbers: in a long series of experiments the fraction of outcomes in which a occurs is approximately p, and the approximation gets better and better as the number of experiments increases. For a regular Markov chain the components of the vector w play the analogous role. That is, the fraction of times that the chain is in state a_i approaches w_i, no matter how one starts.

Example 1 continued: Let us take $p^{(0)} = (.1, .9)$ and see how $p^{(n)}$ changes. Using P as in the example above, we have that $p^{(1)} = (.5167, .4833)$, $p^{(2)} = (.5861, .4139)$, and $p^{(3)} = (.5977, .4023)$. Recalling that $w = (.6, .4)$, we see that these vectors do approach w.

Example 2: As an example let us derive the formulas for the fixed point of a 2×2 transition matrix with positive components. Such a matrix is of the form

$$P = \begin{pmatrix} 1-a & a \\ b & 1-b \end{pmatrix},$$

where $0 < a < 1$ and $0 < b < 1$. Since P is regular, it has a unique probability-vector fixed point $w = (w_1\ w_2)$. Its components must satisfy the equations

$$w_1(1 - a) + w_2 b = w_1,$$

$$w_1 a + w_2(1 - b) = w_2.$$

Each of these equations reduces to the single equation $w_1 a = w_2 b$. This single equation has an infinite number of solutions. However, since w is a probability vector, we must also have $w_1 + w_2 — 1$, and the new equation gives the point $[b/(a + b), a/(a + b)]$ as the unique fixed-point probability vector of P.

Example 3: Suppose that the President of the United States tells person A his intention either to run or not to run in the next election.

Then A relays the news to B, who in turn relays the message to C, and so on, always to some new person. Assume that there is a probability $p > 0$ that any one person, when he gets the message, will reverse it before passing it on to the next person. What is the probability that the nth man to hear the message will be told that the President will run? We can consider this as a two-state Markov chain, with states indicated by "yes" and "no." The process is in state "yes" at time n if the nth person to receive the message was told that the President would run. It is in state "no" if he was told that the President would not run. The matrix P of transition probabilities is then

$$\begin{array}{c} \\ \text{yes} \\ \text{no} \end{array}\begin{array}{c} \begin{array}{cc} \text{yes} & \text{no} \end{array} \\ \begin{pmatrix} 1-p & p \\ p & 1-p \end{pmatrix} \end{array}.$$

Then the matrix P^n gives the probabilities that the nth man is given a certain answer, assuming that the President said "yes" (first row) or assuming that the President said "no" (second row). We know that these rows approach w. From the formulas of the last example, we find that $w = \left(\frac{1}{2}, \frac{1}{2}\right)$. Hence the probabilities for the *nth* man's being told "yes" or "no" approach $\frac{1}{2}$ independently of the initial decision of the President. For a large number of people, we can expect that approximately one-half will be told that the President will run and the other half that he will not, independently of the actual decision of the President.

Suppose now that the probability a that a person will change the news from "yes" to "no" when transmitting it to the next person is different from the probability b that he will change it from "no" to "yes." Then the matrix of transition probabilities becomes

$$\begin{array}{c} \\ \text{yes} \\ \text{no} \end{array}\begin{array}{c} \begin{array}{cc} \text{yes} & \text{no} \end{array} \\ \begin{pmatrix} 1-a & a \\ b & 1-b \end{pmatrix} \end{array}.$$

In this case $w = [b/(a + b), a/(a + b)]$. Thus, there is a probability of approximately $b/(a + b)$ that the nth person will be told that the President will run. Assuming that n is large, this probability is independent of the actual decision of the President. For n large we can expect, in this case, that a proportion approximately equal to $b/(a + b)$ will have been told that the President will run, and a proportion $a/(a + b)$ will have been told that he will not run. The important thing to note is that, from the assumptions we have made, it follows that it is not the President but the people themselves who determine the probability that a person will be told "yes" or "no," and the proportion of people in the long run that are given one of these predictions.

Example 4: The first approximation treated in earlier example leads to a two-state Markov chain, and the results are similar to those obtained in Example 1 above. The second approximation led to a four-state Markov chain with transition probabilities given by the matrix

$$\begin{array}{c} \\ \text{RR} \\ \text{DR} \\ \text{RD} \\ \text{DD} \end{array} \begin{array}{c} \begin{array}{cccc} \text{RR} & \text{DR} & \text{RD} & \text{DD} \end{array} \\ \begin{pmatrix} 1-a & 0 & a & 0 \\ b & 0 & 1-b & 0 \\ 0 & 1-c & 0 & c \\ 0 & d & 0 & 1-d \end{pmatrix} \end{array}.$$

If a, b, c, and d are all different from 0 or 1, then the square of the matrix has no zeros, and hence the matrix is regular. The fixed probability vector is found in the usual way and is

$$\left(\frac{bd}{bd+2ad+ca}, \frac{ad}{bd+2ad+ca}, \frac{ad}{bd+2ad+ca}, \frac{ca}{bd+2ad+ca}\right).$$

Note that the probability of being in state RD after a large number of steps is equal to the probability of being in state DR. This shows that in equilibrium a change from R to D must have the same probability as a change from D to R.

From the fixed vector we can find the probability of being in state R in the far future. This is found by adding the probability of being in state RR and DR, giving

$$\frac{bd+ad}{bd+2ad+ca}.$$

Notice that, to find the probability of being in state R on the election preceding some election far in the future, we should add the probabilities of being in states RR and RD. That we get the same result corresponds to the fact that predictions far in the future are essentially independent of the particular period being predicted. In other words, the process is acting as if it were in equilibrium.

Exercises

1. Which of the following matrices are regular?

(a) $\begin{pmatrix} \frac{1}{2} & \frac{1}{2} \\ \frac{1}{2} & \frac{1}{2} \end{pmatrix}$.

(b) $\begin{pmatrix} 0 & 1 \\ \frac{1}{4} & \frac{3}{4} \end{pmatrix}$. [*Ans.* Regular.]

(c) $\begin{pmatrix} 1 & 0 \\ \frac{1}{3} & \frac{2}{3} \end{pmatrix}$.

(d) $\begin{pmatrix} \frac{1}{5} & \frac{4}{5} \\ 1 & 0 \end{pmatrix}$. [*Ans.* Regular.]

(e) $\begin{pmatrix} \frac{1}{2} & \frac{1}{2} \\ 0 & 1 \end{pmatrix}$.

(f) $\begin{pmatrix} 0 & 1 \\ 1 & 0 \end{pmatrix}$. [*Ans.* Not regular.]

(g) $\begin{pmatrix} \frac{1}{2} & \frac{1}{2} & 0 \\ 0 & \frac{1}{2} & \frac{1}{2} \\ \frac{1}{3} & \frac{1}{3} & \frac{1}{3} \end{pmatrix}$.

(h) $\begin{pmatrix} \frac{1}{3} & 0 & \frac{2}{3} \\ 0 & 1 & 0 \\ 0 & \frac{1}{5} & \frac{4}{5} \end{pmatrix}$. [*Ans.* Not regular.]

2. Show that the 2 × 2 matrix

$$P = \begin{pmatrix} 1-a & a \\ b & 1-b \end{pmatrix}$$

is the regular transition matrix if and only if either

(i) $0 < a \le 1$ and $0 < b < 1$; or

(ii) $0 < a < 1$ and $0 < b \le 1$.

3. Let P be a transition matrix in which all the entries that are not zero have been replaced by x's. Devise a method of raising such a matrix to powers in order to check for regularity. Illustrate your method by showing that

$$P = \begin{pmatrix} 0 & 1 & 0 \\ 0 & 0 & 1 \\ \frac{1}{2} & \frac{1}{2} & 0 \end{pmatrix}$$

is regular.

4. Use the method developed in Exercise 3 to test the following matrix for regularity:

$$P = \begin{pmatrix} 0 & 0 & \frac{1}{10} & 0 & \frac{9}{10} \\ 1 & 0 & 0 & 0 & 0 \\ 0 & \frac{1}{6} & \frac{1}{2} & \frac{1}{3} & 0 \\ 0 & 0 & 0 & 0 & 1 \\ 0 & \frac{3}{7} & 0 & \frac{4}{7} & 0 \end{pmatrix}.$$

5. (a) Give a probability theory interpretation to the condition of regularity.

   (b) Consider a Markov chain such that it is possible to go from any state a_i to any state a_j and such that p_{kk} is not 0 for at least one state a_k. Prove that the chain is regular. [*Hint:* Consider the times that it is possible to go from a_i to a_j via a_k.]

6. Find the fixed point for the matrix in Exercise 2 for each of the cases listed there. [*Hint:* Most of the cases were covered in the text above.]

7. Find the fixed point w for each of the following regular matrices:

   (a) $\begin{pmatrix} \frac{1}{3} & \frac{2}{3} \\ \frac{5}{6} & \frac{1}{6} \end{pmatrix}.$ [*Ans.* $\frac{5}{9}, \frac{4}{9}$.]

   (b) $\begin{pmatrix} .37 & .63 \\ .63 & .37 \end{pmatrix}.$

   (c) $\begin{pmatrix} \frac{3}{8} & \frac{5}{8} \\ \frac{3}{8} & \frac{5}{8} \end{pmatrix}.$

(d) $\begin{pmatrix} \frac{1}{2} & \frac{1}{4} & \frac{1}{4} \\ \frac{1}{3} & \frac{2}{3} & 0 \\ 0 & \frac{1}{4} & \frac{3}{4} \end{pmatrix}$. [*Ans.* $\left(\frac{2}{7}, \frac{3}{7}, \frac{2}{7}\right)$.]

(e) $\begin{pmatrix} \frac{5}{7} & \frac{2}{7} \\ \frac{1}{9} & \frac{8}{9} \end{pmatrix}$.

8. Let $p^0 = \left(\frac{1}{2}, \frac{1}{2}\right)$ and compute $p^{(1)}$, $p^{(2)}$, and $p^{(3)}$ for the matrices in Exercises 7(a), (b), and (c). Do they approach the fixed points of these matrices?

9. Consider the two-state Markov chain with transition matrix

$$P = \begin{array}{c} \\ a_1 \\ a_2 \end{array} \begin{array}{c} \begin{array}{cc} a_1 & a_2 \end{array} \\ \begin{pmatrix} 0 & 1 \\ 1 & 0 \end{pmatrix} \end{array}.$$

What is the probability that after n steps the process is in state a_1, if it started in state a_2? Does this probability become independent of the initial position for large n? If not, the theorem of this section must not apply. Why? Does the matrix have a unique fixed-point probability vector?

10. Compute the first five powers of the matrix

$$P = \begin{pmatrix} .7 & .3 \\ .3 & .7 \end{pmatrix}.$$

From these, guess the fixed-point vector w. Check by computing what w is.

11. Prove that, if a regular 3×3 transition matrix has the property that its column sums are 1, its fixed-point probability vector is $\left(\frac{1}{3}, \frac{1}{3}, \frac{1}{3}\right)$. State a similar result for $n \times n$ transition matrices having column sums equal to 1.

12. The Land of Oz is blessed by many things, but not good weather. They never have two nice days in a row. If they have a nice day they are just as likely to have snow as rain the next day. If they have snow (or rain), they have an even chance of having the same the next day. If there is a change from snow or rain, only half of

the time is this a change to a nice day. Set up a three-state Markov chain to describe this situation. Find the long-range probability for rain, for snow, and for a nice day. What fraction of the days does it rain in the Land of Oz?

[*Ans.* The probabilities are: nice, $\frac{1}{5}$; rain, $\frac{2}{5}$; snow, $\frac{2}{5}$.]

13. A professor tries not to be late for class too often. If he is late one day, he is 95 per cent sure to be on time next time. If he is on time, then the next day there is a 25 per cent chance of his being late. In the long run, how often is he late for class?

14. Consider the three-state Markov chain with transition matrix

$$P = \begin{pmatrix} \frac{1}{4} & \frac{1}{2} & \frac{1}{4} \\ \frac{2}{5} & \frac{3}{5} & 0 \\ 1 & 0 & 0 \end{pmatrix}.$$

(a) Show that the matrix has a unique fixed probability vector.

[*Ans.* $\left(\frac{2}{5}, \frac{1}{2}, \frac{1}{10}\right)$.]

(b) Approximately what is the entry in the third column of the first row of P^{100}?

(c) What is the interpretation of the entry estimated in (b)?

15. A carnival man moves a pea among three shells, A, B, and C. Whenever the pea is under A, he moves it with equal probability to A or B. When it is under B, he is sure to move it to C. When it is under C, he is sure to put it next time under C or B, but is twice as likely to put it under C as B.

Set up a Markov chain taking as states the letters of the shells under which the pea appears after a move. Give the matrix of transition probabilities. Assume that the pea is initially under shell A. Which of the following statements are logically true?

(a) After the first move, the pea is under A or B.

(b) After the second move, the pea is under shell B or C.

(c) If the pea appears under B, it will eventually appear under A again if the process goes on long enough.

(d) If the pea appears under C, it will not appear under A again.

[*Ans.* (a) and (d) are logically true.]

16. In Exercise 15, assume that when the pea is under C, the carnival man is sure to put it next time under C or A, but twice as likely to put it under C as A. If you arrive on the scene after he has been playing for a long time, and bet even money that next time it will turn up under a certain shell, which shell should you bet on,

    (a) Given that you have not seen the previous play?

    (b) Given that the last time the pea was under A?

    Which of the above bets would be fair?

17. Let P be the matrix

$$P = \begin{pmatrix} 1 & 0 \\ \frac{1}{2} & \frac{1}{2} \end{pmatrix}.$$

Compute the unique probability-vector fixed point of P, and use your result to prove that P is not regular.

18. Show that the vector given in Example 4 is the fixed vector of the transition matrix.

19. Show that the matrix

$$P = \begin{pmatrix} 1 & 0 & 0 \\ \frac{1}{2} & 0 & \frac{1}{2} \\ 0 & 0 & 1 \end{pmatrix}$$

has more than one probability-vector fixed point. Find the matrix that P^n approaches, and show that it is not a matrix all of whose rows are the same.

20. A businessman goes to a convention in Chicago once a year. While there, he stays at one of four hotels. Two of them, hotels 1 and 2, are expensive. The other two are very expensive. Assuming that he goes to a given hotel one year, the hotel he goes to the next year is determined by the following matrix of probabilities:

$$\begin{array}{c} \\ 1 \\ 2 \\ 3 \\ 4 \end{array} \begin{array}{c} \begin{array}{cccc} 1 & 2 & 3 & 4 \end{array} \\ \begin{pmatrix} \frac{1}{4} & 0 & \frac{3}{4} & 0 \\ \frac{2}{5} & 0 & \frac{3}{5} & 0 \\ 0 & \frac{1}{2} & 0 & \frac{1}{2} \\ 0 & \frac{1}{3} & 0 & \frac{2}{3} \end{pmatrix} \end{array}.$$

(a) If he stays at hotel 1 one year, what is the probability that he stays in very expensive hotels for at least two of the next three conventions?

(b) Find the long-run probabilities for staying in each of the hotels.

21. For Exercise 20, compute the following:

(a) Given that in 1970 and 1973 he stayed in hotel 1, what is the probability that he stayed in a very expensive hotel during either 1971 or 1972?

(b) If in 1970 he stayed in hotel 1 and in 1972 he was in a very expensive hotel, what is the probability that in 1973 he stayed in an expensive hotel?

[*Ans.* $\frac{7}{18}$.]

22. A professor has three pet questions, one of which occurs on every test he gives. The students know his habits well. He never uses the same question twice in a row. If he used question 1 last time, he tosses a coin, and uses question 2 if a head comes up. If he used question 2, he tosses two coins and switches to question 3 if at least one comes up heads. If he used question 3, he tosses three coins and switches to question 1 if at least one comes up heads. In the long run, which question does he use most often and how frequently is it used?

23. In some cases it makes sense to form a new Markov chain from an old one by condensing two or more states into one.

(a) Show that this can be done for the Land of Oz example (Exercise 12), using as the two new states nice and bad (rain or snow). Set up the matrix of transition probabilities and compute the fixed vector.

[*Partial Ans*: $\left(\frac{1}{5}, \frac{4}{5}\right)$.]

(b) Compare the fixed vector obtained in part (a) to that obtained in Exercise 12.

(c) Would it make sense to condense states in the hotel example (Exercise 20), using as new states expensive (1 or 2) and very expensive (3 or 4)? Explain your answer.

[*Partial Ans:* No.]

24. A certain company decides each year to add a new workers to its payroll, to remove b workers from its payroll, or to leave its workforce unchanged. There is probability $\frac{3}{4}$ that the action taken in the given year will be the same as the action taken in the previous year. The president of the company has ruled that they should never fire workers the year after they added some, and that they should never hire workers the year after they fired some. Moreover, if no workers were added or fired in the previous year, the company is twice as likely to add workers as to fire them.

    (a) Set up the problem as a Markov chain with three states.

    (b) Show that it is regular.

    (c) Find the long-run probability of each type of action.

    (d) For what values of a and b will the company tend to increase in size? To decrease? To stay the same?

[*Ans.* (d) $a > b/2$; $a < b/2$; $a = b/2$.]

Computer Programming

Introduction

Modern high-speed computers have made all forms of computation vastly easier. Calculations that used to take several days to complete can now be carried out in a few seconds. The availability of a modern computer can take a great deal of drudgery out of mathematical computations and makes possible large-scale computations that would otherwise be impossible. This is particularly true in the area of finite mathematics, since each of the branches of mathematics introduced in this book is well suited to computer applications.

It is the purpose of this chapter to give a first introduction to the use of high-speed computers. A computer is an electronic device designed to carry out arithmetical operations and to follow a long list of instructions as to what calculations should be carried out. A computer does no more and no less than a human user instructs it to do; however, it can carry out tasks at tremendous speed and with great accuracy. The key to the use of a computer is learning how to write a set of instructions. Such a set of instructions is called a *programme,* and the art of writing such instructions is known as *programming.* This chapter will give a number of examples of programmes for high-speed computers designed to carry out calculations in finite mathematics. Although only elementary programming techniques will be illustrated, they will be sufficient to carry out many significant mathematical tasks.

The chapter is written so that it may profitably be studied without having a computer available. However, being able to try out examples on a computer will significantly improve the learning experience. Each section will include many exercises that do not require the use of the computer and also some exercises that must be completed on a computer.

In order to communicate with a human being, it is necessary to understand the language he speaks. Similarly, a user must learn a suitable language for communicating with a computer. Fortunately, there are several easily learned languages that most computers "speak." One language widely used, particularly in educational uses, is BASIC. The present chapter provides a brief introduction to the language BASIC. The reader interested in more sophisticated applications, including many further applications to finite mathematics, is referred to the book *Basic Programming*.

Since the language BASIC is almost self-explanatory, it is simplest to learn it by looking at some actual programmes. The programme Example 1 is designed to compute several factorials—specifically, 4!, 6!, and 10!. The programme consists of five lines, each of which contains one instruction for the computer. It will be noted that each line starts with a line number. These numbers are required in BASIC to make it easy to enter corrections to a given programme. For example, if a user wishes to correct a given line, he simply retypes that line with the same line number and the correction is automatically made by the computer. Or, if it is desired to insert a line between, say, lines 20 and 30, one may type the new instruction with any number between 20 and 30 and a correction is automatically made. Thus, it is good practice to choose line numbers with gaps between them (for example, multiples of 10) to allow for the insertion of additional instructions.

Most of the variables in this chapter are represented by capital letters, since many computer terminals print only capitals.

Look now at Example 1. Line 10 in Example 1 instructs the computer to compute 4!, i.e., $1 \times 2 \times 3 \times 4$. To avoid the ambiguity between the dot as a multiplication sign and as a decimal point, BASIC uses an asterisk (∗) to indicate multiplication. Specifically, the LET command in line 10 tells the machine to compute $1 \times 2 \times 3 \times 4$ and to call the answer "X." Thus, after the computer has carried out the instruction in line 10, X will equal 24. This quantity may now be used in the rest of the programme. In line 20 the computer is asked to take X and multiply it by 5 and then by 6, and to call the result Y. Thus, Y will equal 6! or 720. Similarly, on line 30 the previous result is multiplied by 7, 8, 9, and 10, thus obtaining 10!, and calling the result Z. The instruction LET is designed to carry out a wide variety of computations. The format for the LET command is always to put on the right-hand side of the equals sign the computation that is to be carried out and to indicate the name of the result on the left-hand side of the equals sign.

Computations are useless unless the user can see the result. In a long programme, there are many partial results that are of no interest to the user and that would take too long to be typed, so that we don't want to print every result. Therefore there is an instruction in BASIC called PRINT which tells the computer to print or type only the desired results. Line 40 instructs the computer to type out X, Y, and Z. It is up to the programmer to remember that the results stand for 4!, 6!, and 10!, respectively.

EXAMPLE 1:

```
10 LET X = 1*2*3*4
20 LET Y = X*5*6
30 LET Z = Y*7*8*9*10
40 PRINT X,Y, Z
50 END
READY
RUN
```

EXAMPLE 1:

```
24                    720                    3623800

0.062 SEC.
READY
```

The final instruction, in line 50, is END. In all programmes the last instruction must be an END statement. This line both indicates the physical end of the programme and tells the computer to stop.

Immediately after the programme we show the results that are printed as the programme is executed or "RUN." Three numbers are printed which are the desired results. It is important to note that the computations took only a small fraction of a second.

Many interesting and useful programmes can be written with the minimal vocabulary of LET, PRINT, and END. A more sophisticated example is shown in Example 2. Line 10 carries out a subtraction and an addition. Line 20 shows one decimal fraction being divided by the sum of two other decimal fractions. Note that parentheses are inserted as usual. Line 30 requires an additional word of explanation. One usually communicates with a computer through a typewriter like terminal device, and this imposes certain limitations on the way formulas are typed. Specifically, each formula must be contained on a single line. This was already illustrated by the form of division on line 20. Since it is not possible to type an exponent on a higher line, an upward arrow (↑) is used to indicate an exponent. Thus, line 30 asks the computer to raise the number 2.15 to the sixth power and to let the answer be Z.

Line 40 illustrates some additional options available for the PRINT instruction. In Example 1 a comma (,) separated the variables, which is the signal to BASIC to line up the answers in predetermined columns (also called fields), normally up to five columns per line.

EXAMPLE 2:

```
10 LET X = 397-128+511
20 LET Y = .57/(.23+.82)
30 LET Z = 2.15↑6
40 PRINT X; Y; Z; 2*X; Y*Z
50 END
READY

RUN
```

EXAMPLE 2:

```
780  0.542357  98.7713  1560  53.6187

0.059 SEC.
READY
```

If one is not interested in a special format but would simply like to have answers printed one after the other, the answers are separated by semicolons (;), as shown in Example 2. This example also shows that computation instructions may take place within a PRINT statement. In addition to printing X, Y, and Z, we have also asked the computer to PRINT 2*X and Y*Z. Recall that an asterisk (*) is used to denote multiplication to the computer. The results are again shown.

These examples illustrate the fact that once the user masters a simple language for entering requests to the computer, all the hard work can be left to the machine. One of the nice features of modern computers is the fact that one can become quite expert in their use without necessarily having any understanding of how computers work. This is similar to the fact that millions of people use telephones and drive automobiles without having any understanding of the nature of telephone-switching networks or of automobile engines. That is why in this chapter we are concentrating entirely on the art of programming and not on the operation of computers. The following sections will introduce, step by step, some more powerful commands in the language BASIC which will enable the reader to use the computer for more complex calculations.

Exercise

Only Exercises 7-11 require the use of a computer.

1. Write a programme that will compute and print the sum of the first three positive integers, of the first five positive integers, and of the first ten positive integers.
2. Write a programme to compute the cube root of 100 (You will need to recall that a cube root is the same as the $\frac{1}{3}$ power).
3. Write a two-instruction programme to compute $\frac{7}{3}$.
4. The following programme contains three *illegal* instructions, that is, instructions not satisfying the rules we have prescribed. Identify them.

```
10 LET X = .12345/.54321
20 LET Y = X↑Z
30 LET U = XY
40 LET X = X–X
50 PRINT X, Y, U + Z
60 PRINT X, Y, Z²
70 END
80 LET X = Y + Z
```

5. Without using a computer, figure out what would be printed when the following programme is run.

```
10 LET X = 2
20 LET Y = 7-4
30 LET Z = Y↑X
40 LET Z = Z–2*Y+X
50 PRINT Z/X
60 END
```

[*Ans.* 2.5.]

6. Without using a computer, figure out what would be printed when the following programme is run.

```
10 LET X = 20/5
20 LET Y = X↑(1/2)
30 LET Z = 1*2*3
40 LET U = Z–3*Y
50 PRINT U
60 END
```

7. Try the programme of Exercise 4 on a computer to see what error messages are printed.
8. Run the programme of Exercise 2 on a computer. What is the cube root of 100? [*Ans.* 4.64159.]
9. Use a computer to compute (.54321/.12345)*(40/37)↑3.
10. Using a computer, employ a trial-and-error method to find the smallest integer whose fifth power is greater than 1,000,000.
11. Use a computer to compute $\binom{20}{7}$ [*Ans.* 77520.]

More on Basic Language

The programmes in preceding section are not typical in that each instruction is carried out only once by the computer. Such calculations could easily be done with a desk calculator. To make the best use of the great speed of high-speed computers it is desirable to give short programmes which result in hundreds, thousands, or even millions of computer operations. One technique for this is the application of the same instructions to many different sets of data. This will be illustrated in the present section.

Let us suppose that we wish to carry out a number of divisions. Instead of writing a separate instruction for each operation, we can write a single instruction and use it over and over again, as shown in the programme DI-VIDE. Line 20 instructs the computer to PRINT the numbers A, B and their quotient. The trick is to specify various pairs A, B. This is accomplished by storing on line 40 a set of data and instructing the computer on line 10 to pick off two of these numbers. The READ statement instructs the computer to pick the next two numbers on the DATA line and call the first number A and the second number B.

Thus, the first time line 10 is executed, A will equal 12 and B will equal 4, and thus on line 20 this pair of numbers will be printed as will their quotient A/B = 3. The next time line 10 is executed, A will equal 144 and B will equal 12. This will continue until all the data has been used up.

After reading a pair of numbers and printing the result, we would like the computer to go back and do the same two instructions over again. This is accomplished by a GOTO statement. Line 30 instructs the machine to GOTO 10—that is, to go to line 10, which in this case happens to be the beginning of the programme. This is another important use of line numbers.

DIVIDE:

```
10 READ A, B
20 PRINT A, B, A/B
30 GOTO 10
40 DATA 12,4, 144, 12, 10,3.45., 19782*345
50 END
READY

RUN
```

DIVIDE:

```
12          4          3
144         12         12
10          3.45       2.89855
19782       345        57.3391
OUT OF DATA AT 10

STOP
0.079 SEC
READY
```

```
40 DATA 169, 13, 2.97., 1.23, –200, –50, 12345, 1289
RUN
```

DIVIDE:

```
169            13             13
2.97           1.23           2.41463
–200           –50            4
12345          1289           9.57719
OUT OF DATA AT 10

STOP
0.080 SEC.
READY

5 PRINT "FIRST NO.", "SECOND NO.", "QUOTIENT"

RUN
```

DIVIDE:

```
FIRST NO.      SECOND NO.     QUOTIENT
169            13             13
2.97           1.23           2.41463
–200           –50            4
12345          1289           9.57719
OUT OF DATA AT 10

STOP
0.090 SEC.
READY
```

We include with a listing of the programme the results that are obtained. It will be noted that the four pairs of numbers are printed on separate lines with the quotient in each case printed in the third column. This run also illustrates that there are two different ways of terminating a computer programme. One is by reaching an END instruction; the other is for some condition to occur under which the computer can no longer proceed. In this particular case the fifth time it is asked to READ numbers A and B, it finds that there are no numbers left and therefore it prints the OUT OF DATA message. This is a perfectly legitimate way of terminating a programme.

The advantage of writing a programme with READ and DATA statements is twofold. First, it shortens the programme significantly. Second, if the user wishes to reuse the programme with different pairs of numbers, he only has to change line 40 and the rest of the programme is still valid. If one retypes line 40 with different data and types "RUN" again, the new results will be obtained. This is shown as a second RUN of the programme DIVIDE.

We would like to illustrate one more capability of the PRINT instruction.

It is often convenient to label the output of a computer programme. A PRINT instruction will PRINT any label contained between quotation marks exactly as you typed it. We can add labels to the programme DIVIDE as follows:

5 PRINT "FIRST NO.", "SECOND NO.", "QUOTIENT".

Making the line number "5" indicates to the computer that the instruction should be inserted at the beginning of the programme (i.e., before line 10). The three labels will be printed exactly as indicated. The fact that the labels are separated by commas indicates to the machine that they should be typed in separate columns and they will automatically line up with the three columns of output. A new RUN is shown. Such labels may be inserted anywhere in a PRINT statement, as will be seen in the next programme.

A simple computer programme will allow us to convert a probability to odds. We recall that if the probability that an event will occur is P, and $Q = 1 - P$, then the odds in favour of the event may be expressed as P/Q to 1. This is carried out in the programme ODDS.

A set of probabilities is provided on line 90. Line 10 reads one of these probabilities and calls it P. Line 15 computes Q. Line 20 does double duty, both computing the odds and printing the answers.

ODDS:

```
5 PRINT   "PROBABILITY", "ODDS"
10 READ P
15 LET Q = 1–P
20 PRINT P,P/Q;"T0 1"
30 GOTO 10
```

```
90 DATA .5, .75, .6, .3333333, .1
99 END
READY

RUN

ODDS

PROBABILITY          ODDS
0.5                  1 TO 1
0.75                 3 TO 1
0.6                  1.5 TO 1
0.333333             0.5 TO 1
0.1                  0.111111 TO 1
OUT OF DATA AT 10

STOP
0.083 SEC.
READY
```

Note that in line 20 P is followed by a comma so that the probability will occur in one column and the odds in a separate column. After P/Q we have inserted a semicolon (;) so that the quotient is immediately followed by the phrase "TO 1." The effect of this PRINT format is clearly shown in the RUN. Line 30 simply instructs the programme to go back and carry out the computations for the next probability.

As we look at the output we notice that while the first three lines look very clear, the last two are somewhat unnatural. One does not usually say that the odds are 0.5 to 1 in favour of an event; rather one would prefer to say that the odds are 1 to 2 in favour, or 2 to 1 against the event. To achieve this one must have one output format if the odds are in favour of the event (i.e., P greater than Q), and another format if they are not. We must be able to tell the computer that if a certain relationship holds then one thing should happen, and that otherwise something else should happen. This is provided for in BASIC by the IF... THEN instruction.

In the programme ODDS2 we have inserted a test at line 17.

ODOS2:

```
5 PRINT "PROBABILITY", "ODDS"
10 READ P
15 LET Q = 1–P
```

```
17 IF P<Q THEN 40
20 PRINT P,P/Q;"T0 1"
30 GOTO 10
40 PRINT P, "1 TO";Q/P
50 GOTO 10
90 DATA .5,.75,.6,.3333333,.1
99 END
READY

RUN

0DDS2

PROBABILITY          ODDS
0.5                  1 TO 1
0.75                 3 TO 1
0.6                  1.5 TO 1
0.333333             1 TO 2.
0.1                  1 TO 9
OUT OF DATA AT 10

STOP
0.090 SEC.
READY
```

If P is less than Q then the computer is instructed to skip to line 40 and use the alternate output format. But if P ≥ Q the computer goes on to line 20. On line 40 we compute the odds as 1 to Q/P rather than the form used on line 20. A RUN of the modified programme is shown and the reader will note that the odds are now in both a simpler and a more natural form.

The significance of the IF ... THEN statement is that the computer can be instructed to go in one of two different directions. And where it goes depends on the result of previous computations. In those cases where P turns out to be greater than or equal to Q, the computer proceeds with lines 20 and 30. However, if P is less than Q then the computer skips to lines 40 and 50. Thus, the same computer programme can handle both cases, and uses a simple test to distinguish between them.

The general form of this instructions is:

IF [relationship] THEN [line number].

The line number may be any line number in the programme. For the relationship we may use six relational symbols: = (equals), < (is less than), > (is greater than), < = (is less than or equal to), > = (is greater than or equal to), and < > (is not equal to). A more complex example is the following:

IF (X∗Y + 3)< =Z THEN 35.

If the current value of X∗Y + 3 is less than or equal to the current value of Z, the programme takes line 35 as its next instruction. If not, it will proceed in the normal order.

Exercises

Only Exercises 10-14 require the use of a computer.

1. Write a programme that will READ a list of numbers and compute and print their fifth powers.
2. There are many ways of avoiding the "OUT OF DATA" message. One is to have a dummy number at the end of the DATA (say — 99999) and to have the computer terminate when that number is READ. Modify the programme of Exercise 1 by adding an IF statement so that it will terminate in this manner.
3. Modify the programme ODDS2 so that instead of "1 TO 9" it will print "9 TO 1 AGAINST."
4. Modify the programme ODDS2 to avoid the "OUT OF DATA" message.
5. If the DATA in the programme ODDS contains an illegal probability (i.e., a negative number of a probability greater than 1), the result will be meaningless. Insert a test to make sure that P is between 0 and 1.
6. Write a programme that will read a list of numbers from DATA and find its largest element. You will have to avoid the "OUT OF DATA" termination.
7. Modify the programme of Exercise 6 to find the smallest element.
8. The absolute value Y of a number X may be computed in BASIC by writing LET Y = ABS (X). Design a test to check whether two numbers A and B are within 0.001 of each other.

9. In BASIC, INT(X) is the greatest integer less than or equal to the number X. For example, INT(6.235) = 6, INT(10.999) = 10, INT(15) = 15, and INT(–3.52) = –4. Design a test to check whether an integer X is an even number.
10. By means of the IF . . .THEN statement we can remove the trial-and-error method from Exercise 10, in preceding session. Design and RUN a programme that will READ a number A (A > 0), and find the smallest integer N whose fifth power is greater than A.
11. Try out the programme of Exercise 6 on a computer. Does it work correctly when all the numbers in the DATA are negative?
12. In BASIC, SQR(X) is the square root of X. Use a computer to print a table of square roots for the first ten integers.
13. Modify the programme of Exercise 12 to print the square roots of every fifth number between 100 and 200 (i.e., 100, 105, 110,..., 200).
14. There is a fast computational technique for finding the square root of a number A without using the "built-in" SQR function. One lets X be a guess at the square root. (For example, X = 1 is all right.) Let Y = A/X. If X is the correct square root, then Y = X. If not, one uses the average of X and Y as the next guess, and repeats the process until X and Y differ by less than a predetermined small error—say 0.000001. Write and RUN a programme which carries out this technique. Check the answers by means of SQR.

Loops

Let us return to the problem of computing factorials. To compute 10! it is possible to proceed as in Example 1, or to write a single instruction:

```
LET X = 1*2*3*4*5*6*7*8*9*10.
```

However, this is a nuisance even for 10! and becomes very inconvenient for 25!. It also means that if we wish to compute several different factorials we have to write a different line for each one. We would instead like to write a simple set of instructions which say roughly, "Take the numbers from 1 to 10 and multiply them together." This can be accomplished by the pair of instructions FOR and NEXT.

The heart of the programme FCTRL is contained in the "loop" on lines 30-50. The letter K will consecutively stand for the integers 1 through 10. The letter F will contain all the partial results and will eventually equal 10!. To understand line 40 we must remember that in a LET instruction the computer first computes the right-hand side and then lets the letter on the left equal the result. Thus, F is multiplied by the current value of K and this becomes the new value of F. Line 50 instructs the computer to go on to the next value of K until all ten numbers have been used up.

```
FCTRL:

20 LET F = 1
30 FOR K = 1 TO 10
40 LET F = F*K
50 NEXT K
60 PRINT F
99 END
READY

RUN

FCTRL:

3628800

0.052 SEC.
READY
```

Before starting the loop we must tell the computer what the "initial value" of F should be. In computing a product the initial value must always be 1. If we were computing a sum we would start with 0. After the loop is completed we PRINT the final answer on line 60 and then line 99 instructs the computer to stop.

Line no.	*Result*
20	F = 1
30	K = 1
40	F = 1*1 = 1
50	GOES BACK TO 30
30	K = 2
40	F = 1*2 = 2

```
50      GOES BACK TO 30
30      K = 3
40      F = 2*3 = 6
50      GOES BACK TO 30
        •
        •
        •
        •
        •
        •
30      K = 10
40      F = (362880)*10 = 3628800
50      K EXHAUSTED, DOES NOT GO BACK
60      PRINT VALUE OF 10!
99      STOPS
```

Figure: 3.1

Figure 3.1 shows what actually happens as each step in the computation is performed.

It is easy to modify this programme to compute the factorial of an arbitrary number. In FCTRL2 we first PRINT labels and then READ the number N whose factorial we are trying to compute. In line 30 K now goes from 1 to N. In line 60 we have elected to PRINT both N and its factorial. Line 70 instructs the programme to go back and read the next number. Line 90 contains five different values for N. The RUN shows the factorials of these five numbers.

Two comments are in order concerning the output. First, 20! is a number too large for all of the digits to be printed out. Therefore the computer prints it in "scientific notation." The abbreviation E +18 stands for 10^{18}. In other words, the answer is 2.4329×10^{18}. It is also worth noting that 0! came out to be 1 without any special instruction to the computer. This is one more way of showing that 0! = 1 is the "natural convention."

FCTRL2:

```
5 PRINT "NUMBER","FACTORIAL"
10 READ N
20 LET F = 1
```

```
30 FOR K = 1 TO N
40 LET F = F*K
50 NEXT K
60 PRINT N, F
70 GOTO 10
90 DATA 4, 7, 10, 20, 0
99 END
READY

RUN
```

FCTRL2:

```
NUMBER              FACTORIAL
4                   24
7                   5040
10                  3628800
20                  2.4329 E+18
0                   1
OUT OF DATA AT 10

STOP
0.082 SEC.
READY
```

As our next illustration of loops we shall write a short programme that computes an expected value. We recall that an expected value is computed for an experiment whose possible outcomes are numbers by multiplying the numerical outcome A with the probability P of the outcome for each possible outcome, and adding up the results. This is carried out in the programme EXPECT.

EXPECT:

```
10 LET E = 0
20 FOR K = 1 TO 5
30 READ A, P
40 LET E = E + A*P
50 NEXT K
60 PRINT E
90 DATA 1, .3,2, .2,5, .05,–1., .25,–2, .2
99 END
READY
```

RUN

EXPECT:

0.3

0.059 SEC.
READY

Since the expected value E is computed as a sum, its initial value is set to 0. In the loop of lines 20-50, for each of the five possible outcomes we first read the numerical value A and the probability P. We then add to the previous value of E the quantity A∗P. This will become the new value of E. After the loop is completed (by going through all five cases) E will be the expected value. This is printed by line 60. Note that the variable K acts as a counter only and does not otherwise enter the computation.

We see that the expected value in this simple illustration is 0.3. Of course in this case the answer could have been obtained more easily by hand computation. However, if the number of cases were significantly larger and the numbers were not as nice, the computer would indeed be useful.

Let us now turn to an application for which a computer is indispensable. We shall write a computer programme for the "birthday problem".

The problem was to compute the probability that among *R* peopl there are at least two with the same birthday. The trick was to comput first the probability *Q* that all the birthdays are different, then the probabilit we desire will be *P* = 1 – *Q*. In the programme BIRTHDAY lines 15-4 carry out this computation in five simple instructions. The remaining line are designed to allow us to compute the answer for several differen values of R and to PRINT the answers.

BIRTHDAY:

```
5 PRINT "PEOPLE","PROBABILITY"
10 READ R
15 LET Q–1
20 FOR K = 1 TO R
30 LET Q = Q ∗ C366-K)/365
40 NEXT K
```

```
45 LET P = 1–Q
50 PRINT R,P
60 GOTO 10
90 DATA 10,20,22,23,30,50
99 END
READY

RUN
```

BIRTHDAY:

```
PEOPLE              PROBABILITY
10                  0.116948
20                  0.411438
22                  0.475695
23                  0.507297
30                  0.706316
50                  0.970374
OUT OF DATA AT 10

STOP
0.100    SEC
READY
```

That the results agree (except for the fact that these numbers are rounded to three places in Figure 3.1) is not surprising since they were originally obtained by means of a computer. This is a good example in which a simple computer programme and one-tenth of a second of computer time can save hours of laborious hand calculations. The question is often raised: What is the probability of having more than one coincidence of birthdays in a given group? That is, what are the chances that there will be three people with the same birthday or two pairs of identical birthdays or even larger coincidences? This probability can be computed in two steps. One first computes (as above) the probability of having some kind of coincidence. Then one computes separately the probability of having precisely one pair of people with the same birthday. The difference of these two quantities will give the probability of a multiple coincidence.

BIRTH2:

```
5 PRINT "PEOPLE","PROBABILITY"
10 READ R
```

```
15 LET Q = 1
20 FOR K = 1 TO R
30 LET Q = Q * (366-K)/365
40 NEXT K
45 LET P = 1-Q
50 LET E = 1
60 FOR K = 1 TO R-l
65 LET E = E * C366-K>/365
70 NEXT K
75 LET E = E/365
77 LET E = E*R*(R–l)/2
80 PRINT R,P–E
85 GOTO 10
90 DATA 20,25,30,35,36,40,50
99 END
READY

RUN
```

BIRTH2:

```
PEOPLE          PROBABILITY
20              8.82398 E-2
25              0.189257
30              0.326101
35              0.480722
36              0.511303
40              0.630989
50              0.855524
OUT OF DATA AT 10

STOP
0.134 SEC.
READY
```

The programme BIRTH2 is designed to compute this probability for various numbers of people. The quantities Q and P are computed as before. The quantity E will stand for the probability of exactly one pair with the same birthday. Let us first calculate the probability that the first two people have a specific birthday, say October 26, and that all the other people have different birthdays. This probability is $\frac{1}{365} \times \frac{1}{365} \times \frac{364}{365} \times \frac{363}{365} \times ....$ However, the same two people could have had a coincidence of birthdays on any of 365 days, and therefore we must multiply the answer by 365. This will cancel the first factor of $\frac{1}{365}$. This calculation is carried out in BIRTH2, lines 50-75. We must still correct this answer since we have so far assumed that it is the first two people who have a coincidence of birthdays. Such a coincidence may occur for any pair from among the R people, and therefore we must multiply the answer by $\binom{R}{2} = \frac{R \times (R-1)}{2}$, which is carried out in line 77. You should "step through" the programme BIRTH2 by hand to see that it is actually carrying out the calculations described above.

The programme prints the probability of a multiple coincidence for several different numbers of people. We notice that for 25 people—a number for which we already have a better-than-even chance of having some coincidence—the probability of a greater coincidence is less than .2. The smallest number of people for which a multiple coincidence has better than an even chance is 36. We note that for 50 people the probability of a multiple coincidence is very high.

We have now discussed nine instructions in BASIC. It is significant that these nine instructions are sufficient to write many interesting programmes. They are summarised in Figure 3.2 for the reader's convenience.

Instruction	***Example***	***Purpose***
LET	LET X = 2 + 3	Carries out computations
PRINT	PRINT X.Y.X + Y	Prints results
END	END	Terminates computation
READ	READ A, B	Enters numbers from DATA
DATA	DATA 5, -2, 3.4	Stores data
GOTO	GOTO 20	Transfers programme control
IF... THEN	IF X>3 THEN 20	Performs a test
FOR	FOR N = 1 TO 8	Starts a loop
NEXT	NEXT N	Closes a loop

Figure 3.2: Instructions for nine-word BASIC

Exercises

Only Exercises 9-14 require the use of a computer.

1. Use FOR and NEXT to write a programme that will compute the seventh powers of the first ten positive integers.
2. Write a programme that will compute the cube roots of the integers from 1 to 20.
3. A loop need not run through all the integers specified in the FOR statement. For example, the instruction

   FOR N = 1 TO 15 STEP 2

   will run through the odd numbers from 1 to 15. Write a programme to compute the cube roots of the multiples of 10 from 10 to 100.
4. Write a programme to print the fifth powers of the even integers up to 30.
5. If we know how many numbers there are on the DATA list, we may avoid the OUT OF DATA message by reading the data within a loop.

   Modify DIVIDE in this manner [*Hint:* Remember that a pair of numbers is READ each time].
6. Write a programme to compute the sum of the first 100 integers. What must the initial value of the sum be?
7. Modify the programme of Exercise 6 to READ a number N and then to compute the sum of the first N integers.
8. Write a programme to compute the sum of an arithmetic series. READ only the numbers A, D, N, and have the computer construct the sum of the series with N terms, starting with A, and increasing by D each time, i.e., the series is

   $$A + (A + D) + (A + 2D) + \cdots + (A + (N-1)D).$$
9. RUN the programme of Exercise 7 for several values of N. Check that the answer is always $N(N + 1)/2$.
10. In BASIC, LOG(X) is the natural logarithm of X. Print a table of natural logarithms for the first ten integers.
11. RUN the programme of Exercise 3 on a computer.
12. Write a programme that computes the sum of the first N odd integers. RUN it for several values of N, and guess what the general formula for the sum is.

13. The technique described in Exercise 5 is not the best one, since when the number of DATA elements is changed, the loop must also be changed. This may be avoided by starting DATA with a single number that tells us how many times we have to go through the loop. Say this is N. Then we start our loop with

    FOR K = 1 TO N.

    Modify DIVIDE accordingly, and RUN it.

14. In the Land of Oz the calendar year has 534 days. Modify the programmes BIRTHDAY and BIRTH2 accordingly.
    (a) How many people should we have in order to have a better-than-even chance of a coincidence? [*Ans.* 28.]
    (b) How many for a better-than-even chance of a multiple coincidence?

Lists and Tables

In many applications we wish to work with an entire array of numbers at the same time. For this BASIC provides "lists" and "tables." A list can be used to store a sequence of numbers, while a table contains a two-dimensional array of numbers. We shall see in the next section that lists can also be used as vectors and tables also as matrices, allowing us to carry out matrix operations.

We have had previous arrays of numbers contained in our DATA statement. However, in each case we READ the numbers one or two at a time, and once we made use of them, we could afford to forget them. A list becomes important when the entire array must be remembered. For example, if we wish merely to read a sequence of numbers, multiply each one by 5, and print the results, there is no need to employ a list. However, an application as simple as reading a sequence of numbers and printing them out in the opposite order requires the use of a list. This is shown in the programme BACK.

BACK:

```
10 FOR I = 1 TO 8
20 READ L(I)
30 NEXT I
40 FOR I = 8 TO 1 STEP -1
50 PRINT L(I);
```

```
60 NEXT I
90 DATA 1, 3, 6, 10, 15., 21, 28, 36
99·END
READY

RUN
```

BACK:

```
36    28    21    15    10    6    3    1

0.066  SEC.
READY
```

BASIC allows one list or table for each letter of the alphabet. For example, if the letter L is used to designate a list, then L(3) will stand for the third element of the list, while L(7) will stand for the seventh element. It would be more common mathematical notation to write these as L_3 and L_7 However, these cannot be typed on the devices one uses to communicate with computers. Lists and tables are nonetheless often referred to as "subscripted variables" because of the more usual mathematical notations for them.

We have found it convenient in earlier chapters of this book to refer to an arbitrary element of a list of numbers by a notation such as L_x. The analog in BASIC is to write the formula L(l). Then as I runs through the numbers 1,2,..., the quantity of L(l) will run through the various elements of the list. We take advantage of this possibility in the programme BACK as lines 10-30 read the entire list of eight elements. The first time through the loop I equals 1 and therefore on line 20 we read L(1); thus the data element "1" on line 90 becomes the first element of the list. The second time I = 2 and therefore we read L(2) and thus the data element "3" becomes the second element of the list. Finally, the data element "36" will become L(8).

To print out the list in reverse order we can again employ a three-instruction loop. We want to print each L(I); however, we want I = 8, 7, . . ., 1. Line 40 shows an additional flexibility of the FOR instruction. One can specify any step size by which the programme proceeds (If no step is specified, the computer assumes that the step size is 1). In this case we specify STEP- 1; thus I = 8 the first time, then 7, then 6, etc.

The semicolon (;) on line 50 will assure that the numbers are printed one after the other without any extra space. If instead of a semicolon we had used a comma (,) the numbers would be printed in columns. If we had used no punctuation at the end of the line, each component would have been printed on a new line.

The programme DICE computes the probability of winning in the game of craps. We shall use the list P to store the probabilities for various possible sums when two dice are rolled. For example. P(5) will be the probability of shooting a 5, which we know to be $\frac{4}{36}$. Whenever a list is used in BASIC, space is automatically allocated in the computer for up to ten elements. Similarly for any table, BASIC will allocate for a table of size up to a 10×10. If larger lists and tables are desired, one must specify this through the use of a DIM or dimension statement. Thus, in the programme DICE we indicate the list P will have 12 elements. In case several different uses are contemplated for the same list, one must specify a DIM large enough to accommodate the longest list.

Lines 20-40 set up the probabilities for rolling a 2 through a 7. We leave to the reader the verification of these formulas. Lines 50 through 70 take advantage of the symmetry of the problem; e.g. the probability of an 8 is the same as the probability of a 6. At the end of this loop all the various probabilities for different totals on a single roll have been computed and the next step is to compute the probability W for winning in the game of craps.

You will recall that if a 7 or 11 turns up on the first roll, we win immediately. This is reflected in line 100. To this probability we must add the probability of "making our point." That is, if the initial roll is 4, 5, 6, 8, 9, or 10, then we must keep rolling until we either repeat that number (in which case we win) or until a 7 turns up.

DICE:

```
10 DIM P(12)
20 FOR K = 2 TO 7
30 LET P<K) = (K-I)/36
40 NEXT K
50 FOR K = 8 TO 12
60 LET P(K) = P(14-K)
70 NEXT K
```

```
100 LET W = P(7) + P(ll)
110 FOR K = 4 TO 10
112 IF K=7 THEN 130
115 LET C = P(K)/(P(K)+P(7))
120 LET U = W + P(K)*C
130 NEXT K
170 PRINT W
180 PRINT 244/495
199 END
READY

RUN
```

DICE:

```
0.492929
0.492929

0.076 SEC.
READY
```

This calculation is carried out in the loop on lines 110-130. Since our "point" may be 4, 5, 6, 8, 9, or 10, we allow the loop to run from 4 to 10. However, we must exclude 7 as a possibility and this is the reason for line 112: if K is 7, we jump to the end of the loop—in other words, we eliminate this possibility. Line 115 computes the conditional probability that the number K will be repeated given that we shall get either K or a 7. This is the simplest way of computing the probability of getting a K before we get a 7. On line 120 we add to our previous winning probability the probability that we both have K as our initial point and that we win with it. By the time the loop is completed W will equal the probability of winning at craps. This is printed on line 170. We had calculated this probability earlier as $\frac{244}{495}$ and we also print this quantity for comparison. We note from the RUN that the two answers are identical.

Let us now consider the use of tables. If T stands for a table, we must indicate which row and which column in the table we are looking at. Thus, T(3, 5) will stand for the table entry in row 3 and column 5. As usual the arguments (or subscripts) may be variables. Thus, T(I, J) will stand for the entry in row I and column J, which is more usually indicated by T_{ij}. As an illustration we shall recompute one of the tables previously computed in the book. This will be the table of binomial coefficients, usually known as the Pascal triangle. We shall compute the quantities $\binom{N}{J}$ for the values $N = 0, 1,2,..., 30$ and all possible values of J, namely $J - 0, 1,..., N$. Only two facts are needed to compute the Pascal triangle. One is the fact that $\binom{N}{0} = \binom{N}{N} = 1$. The other is the fact that any entry "inside" the triangle is equal to the sum of the two entries immediately above it.

BINOMC:

```
10 DIM B (30,30)
20 FOR N = 0 TO 30
30 LET B(N, 0) = 1
40 LET B(N, N) = 1
50 FOR J = 1 TO N-l
60 LET B(N, J) = B(N-1, J-1) + B(N-1, J)
70 NEXT J
80 NEXT N
90
100 PRINT " N", "J", "SINOM"
110 FOR K = 1 TO 4
120 READ N, J
130 PRINT N, J, B(N, J)
140 NEXT K
190 DATA 10, 5, 15, 3, 25., 10, 30.. 15
199 END
READY

RUN
```

BINOMC:

N	J	BINOM
10	5	252
15	3	455
25	10	3268760
30	15	1.55118 E+8

0.188 SEC.
READY

In the programme BINOMC, line 10 saves enough space for a 30 × 30 table. The entry B(N, J) will stand for $\binom{N}{J}$. The entire calculation of the triangle is carried out on lines 20-80. We let N run from 0 through 30. For each given N we first fill in the $\binom{N}{0}$ and $\binom{N}{N}$ on lines 30 and 40. Then we start the loop on J in which J runs from 1 through N – 1 to compute the "inside" entries. Line 60 simply states that a given entry of B is the sum of two entries on the previous row. Lines 70 and 80 close the two Loops.

This is our first example of a "double loop." Such a double loop is legal as long as one loop is completely contained within the other one. The interpretation is very simple: the computer picks a first value for N and then runs through all the indicated values of J; it then picks the next value of N and repeats the procedure; and so on.

BINOMPR:

```
10 DIM B(30, 30)
20 FOR N = 0 TO 30
30 LET B(N, 0) = 1
40 LET B(N, N) = 1
50 FOR J = 1 TO N–1
60 LET B(N, J) = B(N–1, J–1) + B(N–1, J)
70 NEXT J
80 NEXT N
90
100 PRINT "N","J","P","PROB."
110 FOR K = 1 TO 3
120 READ N, J, P
130 PRINT N, J, P, B(N, J)*P↑J*(1–P) ↑ (N–J)
140 NEXT K
190 DATA 10, 5, .3, 15, 7, .4, 30, 15, .5
199 END
READY
```

RUN

BINOMPR:

N	J	P	PROB.
10	5	0.3	0.102919
15	7	0.4	0.177084
30	15	0.5	0.144464

0.185 SEC.
READY

Note that in this example therange of the second loop depends on the value of N in the first loop. It is in this way that we fill out a triangle. If we were instead filling out a rectangle or a square, the possible values of J would not depend on the value N.

Just to show that the calculations are correct, we end up by printing four binomial coefficients. It is worth noting that the entire calculation of nearly 500 binomial coefficients—including some very large ones, as can be seen on the last line of the output—took only about two-tenths of a second. The same calculation by paper and pencil is a formidable task.

Some additional comments are in order. Line 90 is blank. This has no effect on the computations, but it separates the two major portions of the programme for easier reading. On line 100 we PRINT appropriate labels. It should be noted that since we use commas both here and on line 130, the outputs automatically line up. The loop on lines 110-150 is employed to avoid the "OUT OF DATA" message.

Once we have binomial coefficients computed, they can be used for the solution of many kinds of problems. As an illustration we have included the programme BINOMPR which computes binomial probabilities. Lines 10-90 are identical with the previous programme since these simply compute the Pascal triangle. In the rest of the programme we read the value of N, J, P, and compute the probability of precisely J successes in N trials with probability P for success on each trial. In line 130 we print N, J, and P and then compute and print the binomial probability. For example, the second line of the output shows that if we have 15 experiments with probability .4 for success on each experiment, then the probability of precisely 7 successes is about .177.

Exercises

1. We wish to read N numbers from a DATA list and perform a task on them. Which of the following tasks require that the numbers be stored in a list?
   (a) Find the largest number.
   (b) Print the even-numbered entries.
   (c) Print first the even-numbered and then the odd-numbered entries.
   (d) Find the sum of the numbers.
   (e) Find the sum of the first and last entries.
   (f) Arrange the numbers in order. *[Am.* (c) and (f).]
2. To use the same programme for tables of different dimensions, one should read first the dimensions of the table, and then read the table (Otherwise one does not know how many rows and columns to read). Write such a programme.
3. Write a programme that will read a table and compute the row sums.
4. Write a programme to read a list of four entries and a list of seven entries and construct a table T so that T(I, J) is the product of the Ith entry of the first list and *J*th entry of the second list.
5. Write a programme to read a list and arrange the numbers in increasing order.
6. Modify the programme of Exercise 5 to arrange the numbers in decreasing order.
7. Use the programme DICE to compute the expected value of the game if $1 is bet each time.
8. Use DICE to compute the expected value of the game if we win $2 on 7 or 11, lose $3 on 2, 3, or 12, and win or lose $1 for making or failing to make our point.
9. Compute the row sums of the Pascal triangle for N = 0, 1,..., 10. Use the binomial theorem to explain the results.
10. For N = 0, 1,. . ., 10 compute the sum of $\binom{N}{J} 2^J$. Use the binomial theorem to explain the results.

11. This is an exercise in modular arithmetic. For I, J = 1, 2,..., 6 let T(I, J) be I*J reduced by 7's. That is, if the product is 7 or greater, keep subtracting 7 until the result is less than 7. Print the table. What pattern do you observe?

Vectors and Matrices

A natural use of lists and tables is to use them for vectors and matrices and to carry out matrix operations with them. BASIC recognises this use by having a special set of instructions that enable one to carry out the matrix operation in a single step. We shall illustrate this by writing two programmes for the addition of vectors, one not using the special instructions and one using them.

In the programme VECADD the calculations are accomplished in four triples of instructions (loops). The first three instructions read a seven-component vector A, the second triple reads a similar vector B, and the third triple of instructions computes the vector sum letting the vector C stand for the answer. Finally, lines 100-120 print the answer.

VECADD:

```
10 FOR I = 1 TO 7
20 READ A(I)
30 NEXT I
40 FOR I = 1 TO 7
50 READ B(I)
60 NEXT I
70 FOR I = 1 TO 7
80 LET C(I) = A(I) + B(I)
90 NEXT I
100 FOR I - 1 TO 7
110 PRINT C(I);
120 NEXT I
190 DATA 1,2,3,4,5,6,7
191 DATA 5, 8, 2, 0, –1., –3, –7
199 END
READY

RUN
```

VECADD:

```
6     10     5     4     4     3     0

0.083   SEC.
READY
```

In the programme VECADD2 we have replaced each triple of instructions (each loop) by a single special instruction. One signals to BASIC that an instruction is a special matrix instruction by starting with "MAT." The first two instructions each read a seven-component vector, the third instruction carries out the vector addition, and the fourth instruction prints the vector.

VECADD2:

```
10 MAT READ A(7)
40 MAT READ B(7)
70 MAT C = A + B
100 MAT PRINT C;
190 DATA 1,2,3,4,5,6,7
191 DATA  5,8,2,0,–1,-3,–7
199 END
READY

RUN
```

VECADD2:

```
6    10    5    4    4    3    0

0.076   SEC.
READY
```

The comparison of the two programmes will show exactly what the MAT instructions accomplish. Clearly the second programme is simpler and shorter. The saving is even greater when we are dealing with a matrix or if we want to do more complicated matrix operations.

The next programme, MATSUB, is similar to VECADD2 except that we are dealing with matrices and we perform a subtraction of two matrices. We have arranged the data for the two 3 × 4 matrices A and B on lines 50-52 and 60-62, respectively.

MATSUB:

```
10 MAT READ A(3, 4)
20 MAT READ B(3, 4)
30 MAT C = A – B
40 MAT PRINT C;
49
```

```
50 DATA 1, 2, 3, 4
51 DATA 5, 6, 7, 8
52 DATA 7, 6, 5, 4
59
60 DATA 4, 3, 2, 1
61 DATA 0, –1, –2, –3
62 DATA –2, –3, –2, –1
69
99  END
READY

RUN
```

MATSUB:

```
–3  –1   1   3
5    7   9  11

0.084 SEC.
READY
```

This was done purely for the convenience of the reader, so that he may easily check the computed answer. One could have listed the data all on one line, or divided it among several lines in an arbitrary way, as long as the data appeared in the order in which the programme calls for it. However, it is usually good practice to arrange the data in a neat format for easier proofreading. The next programme, MATMPY, carries out a matrix multiplication. It reads a 3 × 4 matrix A and a 4 × 2 matrix B and computes the product, a 3 × 2 matrix C = AB. The programme MATMPY should be self-explanatory.

MATMPY:

```
10 MAT READ A (3,4)
20 MAT READ B(4, 2)
30 MAT C = A*B
40 MAT PRINT C;
49
50 DATA 1, 2, 3, 4
51 DATA 5, 6, 7, 8
52 DATA 7, 6, 5, 4
59
60 DATA 2, 1
```

```
61 DATA 3, 2
62 DATA –1, 0
63 DATA –3, –4
69
99 END
READY
RUN
```

MATMPY:

```
-7   -11

-3   -15

15     3

0.079 SEC.
READY
```

The single most powerful instruction in BASIC is the one-line command that inverts a matrix. This is illustrated on line 20 of the programme MATINV. Line 10 reads a 3 × 3 matrix A. In line 20 we let $B = A^{-1}$. We then PRINT the inverse. Naturally, we obtain the same result as in that section.

Since matrix inversion is available, it provides a convenient method of solving N equations in N unknowns. We know that we can write equations in the form AX = B. Here A contains the *n* X *n* matrix of coefficients of the left-hand sides of the equations while B is a vector containing the right-hand sides of the equations. The vector X contains the unknowns. If the matrix A has an inverse, then the solution may be written in the form $X = A^{-1}B$. This is carried out in the programme EQU.

MATINV:

```
10 MAT READ A(3, 3)
20 MAT B = INV(A)
30 MAT PRINT B;
39
40 DATA 1, 4, 3
41 DATA 2, 5, 4
42 DATA 1, –3, –2
49
99 END
READY
```

```
RUN

MATINV:

2.     -1.     1.
8.     -5.     2.
-11.    7.    -3.

0.082 SEC.
READY
```

To make the programme more general, we first read the value of N. This will enable us to solve different numbers of equations in different numbers of unknowns by simply changing the data. We then read the matrix A and the vector B, compute the inverse of A, and compute the solution X on line 50. The answers are printed on line 60. The reader can easily verify that the solution is correct.

```
EQU:

10 READ N
20 MAT READ A(N, N)
30 MAT READ B(N)
40 MAT I = INV(A)
50 MAT X = I*B
60 MAT PRINT X;
69
70 DATA 4
79
80 DATA 4, 2, 6, 8
81 DATA 1, 2, 3, 4
82 DATA 4, 3, 2, 1
83 DATA 8, 6, 2, 4
89
90 DATA -12, -5, 5, 12
99  END
READY

RUN

EQU:

1    2.  -2.  -1.

0.087 SEC.
READY
```

If the reader has ever attempted to solve four equations in four unknowns, he will be happy to see that the same solution may be obtained on a computer in a small fraction of a second. Indeed, the same programme will yield the solution of 50 equations in 50 unknowns in roughly 6 seconds! Although we illustrate programming techniques in terms of very simple examples, it is important to remember that the same techniques work on problems much too large to do by hand and often take only a few seconds to do on the computer.

One word of warning is in order for the last two programmes. Not all matrices have inverses, and therefore one should really insert in the programme a test as to whether the matrix does have an inverse. Such a test exists in BASIC but it is beyond the scope of our present treatment.

Exercises

Only Exercises 8-12 require the use of a computer.

1. Write a programme that will add two matrices without using MAT instructions.
2. Write a programme that will compute the tenth power of a square matrix P.
3. In BASIC, a vector of all l's, say of five components, may be constructed by means of the instruction

   MAT X = CON(5).

   Write a programme to read a 7 × 5 matrix *A* and compute *AX* (where *X* is the vector of all l's). Interpret *AX*.
4. In BASIC a vector or a matrix may be multiplied by a number as follows:

   MAT Y = (5.2)*X.

   Then Y = 5.2X. Write a programme to read two five-component vectors X and Y and to compute

   (a) 3Y.

   (b) X + 2Y.

   (c) 5.2X-3.17Y.
5. Write a programme that will read two 3 × 4 matrices A and B and compute

(a) A-3B.

(b) 5.07A + 7.98B.

6. Write a programme that will read a 3 X 3 matrix A, compute its inverse A^{-1}, and compute the product AA^{-1}.
7. Write a programme that will check for two 4 × 4 matrices A and B whether AB = BA. It should print "YES" if they are equal.
8. Set up DATA for a 3 × 4 matrix A, a 4 × 2 matrix B, and a 2 × 3 matrix C. Compute:

(a) ABC.

(b) BCA.

(c) CAB.

(d) BAC.

9. Set up DATA for a 10 × 10 matrix all of whose components are zeroes or ones. Attempt to invert it *[Hint:* Make sure that no row and no column consists entirely of zeroes].
10. Use the computer to verify that $(AB)^{-1} = B^{-1}A^{1}$.
11. RUN the programme of Exercise 6. How close is the final matrix to an identity matrix? (In general one expects some round-off errors).
12. Try the programme of Exercise 7 for several examples. Can you find examples where AB = BA?

Applications to Markov Chains

Matrix theory of Markov chains states that if P is a regular transition matrix, its powers approach a matrix whose rows are identical, each row being a probability vector with positive components. This can be illustrated by taking such a transition matrix and raising it to higher and higher powers. To speed up the process we shall square the matrix each time, so that we shall compute the powers P^2, P^4, P^s, P^{16}, and p^{32}.

We have arranged the transition matrix so that the first row corresponds to "nice," the next to "rain," and the final row to "snow." We first read the 3 × 3 transition matrix. Then in the loop in lines 15-50 we square this matrix five times. Line 20 carries out the actual squaring and line

30 prints the new matrix. On line 40 we let S take the place of P so that when we go through the loop again it is the new matrix that is squared.

Looking at the output we have a dramatic demonstration of the fundamental theorem.

OZ:

```
10 MAT READ P(3, 3)
15 FOR T = 1 TO 5
20 MAT S = P*P
30 MAT PRINT SJ;
40 MAT P = S
50 NEXT T
90 DATA   0,.5,.5
91 DATA   .25,.5,.25
92 DATA   .25,.25,.5
99 END
READY
RUN
```

OZ:

```
0.25          0.375        0.375
0.1875        0.4375       0.375
0.1875        0.375        0.4375

0.203125      0.398437     0.398437
0.199219      0.402344     0.398437
0.199219      0.398437     0.402344
0.200012      0.399994     0.399994
0.199997      0.400009     0.399994
0.199997      0.399994     0.400009

0.2           0.4          0.4
0.2           0.4          0.4
0.2           0.4          0.4
0.2           0.4          0.4
0.2           0.4          0.4
0.2           0.4          0.4

0.139 SEC.
READY
```

By the third squaring—that is, when we look at P^8—the rows of the matrix are almost identical and have nearly assumed their limiting values. In the next two printed matrices, P^{16} and P^{32}, we see that, to the Accuracy to which results are printed, we have the limiting probabilities of .2 for nice, .4 for rain, and .4 for snow.

We shall next turn to absorbing Markov chains as treated earlier. We shall show how easy it is to compute the basic quantities N, T, and B on a computer.

In order to specify an absorbing chain, we need to know the number of transient states* (K) and the number of absorbing states (L). We need also to specify the two submatrices R and Q. This is accomplished in lines 10 and 20 of the programme TRANS and in the DATA statements. The computation and printing of all the other quantities is accomplished in lines 30-90. Lines 30 and 40 illustrate two additional MAT commands. We can set up an identity matrix of specified size and a constant vector (vector of all l's) in single instructions. We then let D = I $-Q$ and compute the inverse N = $(I - Q)^{-1}$. Similarly, we compute T and B. We have thus translated the three main theorems on absorbing chains into six instructions in BASIC. Finally, on line 90 we print the matrices N, T, and B.

TRANS:

```
10 READ K, L
20 MAT READ R(K, L), Q(K, K)
25
30 MAT I = IDN(K, K)
40 MAT C = CON(K)
50 MAT D = I–Q
60 MAT N = INV(D)
70 MAT T = N*C
80 MAT B = N*R
90 MAT PRINT N, T, B
95
100 DATA 3, 2
105
110 DATA .5, 0
111 DATA 0, 0
112 DATA 0, .5
```

```
115
120 DATA 0, .5, 0
121 DATA .5, 0, .5
122 DATA 0, .5, 0
125
199 END
READY

RUN

TRANS:

1 .5      1         0.5
1         2           1
0.5       1         1.5

3         4           3

0.75  0.25
0.5     0.5
0.25  0.75

0.116 SEC.
READY
```

Therefore we change the data to run a more substantial example. Our large example will be a random walk with ten transient states and with probability .7 of taking a step to the right. This chain is symbolically indicated in Figure 3.3. To find the fundamental matrix for this chain we have to invert a 10 × 10 matrix, and therefore we have a more substantial challenge for the computer. It is important to note that the main part of the programme does not have to be changed at all. The only changes needed are in the DATA statements (If there were more than ten transient states, one would also have to insert a DIM statement). We also elected to omit the printing of the matrix N, since it is very large and not particularly interesting.

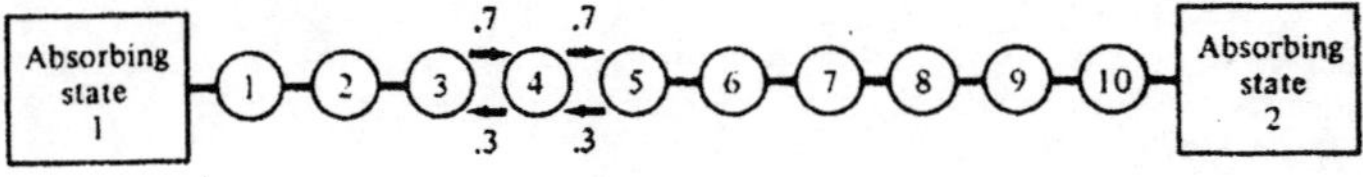

Figure 3.3

We show the new data statements and the run of the programme TRANS2. Looking first at the second half of the output, the matrix B, we note that for most of the states we are almost certain to end up in the second absorbing state (i.e., at the right-hand end). It is surprising that even if one starts in transient state 1, way over on the left, one has a better-than-even chance of ending up on the right. The vector T, containing the expected number of steps before being absorbed, is also quite interesting. If we start near the right-hand endpoint, absorption takes place very fast, as may be expected. However, the result is not obvious on the left-hand side. For example, if one starts in transient state 1, there is probability .3 of being absorbed in a single step. On the other hand, it is more likely that absorption will take place at the right-hand endpoint, which will take a considerable amount of time.

```
100 DATA 10,2
110 DATA .3,0
111 DATA 0,0
112 DATA 0,0
113 DATA 0,0
114 DATA 0,0
115 DATA 0,0
116 DATA 0,0
117 DATA 0,0
118 DATA 0,0
119 DATA 0,.7
120 DATA 0,.7,0,0,0,0,0,0,0,0
121 DATA .3,0,.7,0,0,0,0,0,0,0
122 DATA 0,.3,0,.7,0,0,0,0,0,0
123 DATA 0,0,.3,0,.7,0,0,0,0,0
124 DATA 0,0,0,.3,0,.7,0,0,0,0
125 DATA 0,0,0,0,.3,0,.7,0,0,0
126 DATA 0,0,0,0,0,.3,0,.7,0,0
127 DATA 0,0,0,0,0,0,.3,0,.7,0
128 DATA 0,0,0,0,0,0,0,.3,0,.7
129 DATA 0,0,0,0,0,0,0,0,.3,0
199 END
READY

RUN
```

TRANS2:

```
13.22    17.45    17.84    16.57    14.60
12.33     9.93     7.47     4.99     2.50
```

```
0.4285   0.5715
0.1836   0.8164
0.0786   0.9214
0.0336   0.9664
0.0144   0.9856
0.0061   0.9939
0.0026   0.9974
0.0010   0.9990
0.0004   0.9996
0.0001   0.9999

0.288 SEC.
READY
```

We find that the longest time to absorption, 17.84 steps, is from transient state 3.

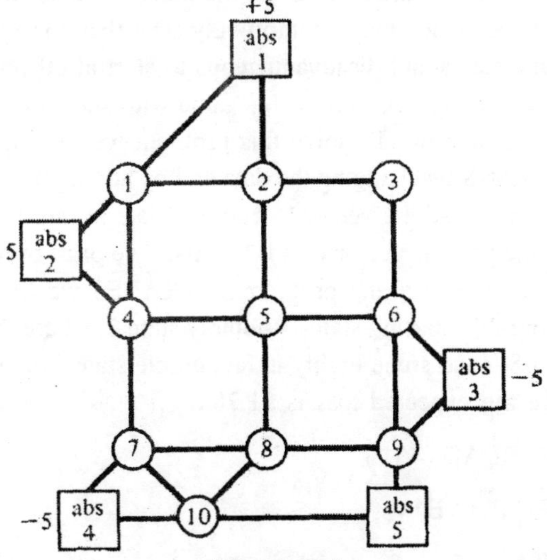

Figure: 3.4

For our final application in this section we consider a "Markov chain game." Any absorbing Markov chain can be turned into a game as follows. First assign a "value" to each absorbing state. A positive value may be interpreted as a prize one wins if one reaches this state while a negative value is a penalty to be paid if that state is reached. A player starts at a given transient state and moves from state to state in accordance with the transition probabilities of the Markov chain until an absorbing state is reached, where he receives a pay-off or pays a penalty. The interesting question is what the expected pay-off is for various different possible starting transient states. If the values are collected into a vector V with as many components as there are absorbing states, then it is very easy to see that the expected pay-off for different starting states is given by the vector BV.

As an illustration we shall consider a game based on a two-dimensional random walk as shown in Figure 3.4. There are ten transient states.

From each of these the player moves to any of the states to which it is connected with equal probabilities. Thus, from state 4 there is probability $\frac{1}{4}$ of moving to states 1, 5, or 7 or to absorbing state 2. There are five absorbing states, and one receives a prize of \$5 at the first one and \$10 at the last one and loses \$5 at the other three absorbing states. The pay-offs balance out, but it is intuitively clear that it is advantageous to start at some states and disadvantageous to start at others. However, if one is offered the chance to play this game with state 5 as the starting state, should one accept? To solve this problem we have modified the programme TRANS by changing the data and by adding three additional lines. Line 25 will read the vector V from the data in line 130. And line 81 computes the pay-off vector P as B*V. Also, we print only the vector P. When we run the resulting programme TRANS3 we find that there are some favourable starting states—notably state 2, where the expected pay-off is \$1.08—and some highly unfavourable states, the worst being state 4, where the expected loss is \$1.78.

```
25 MAT  READ  V(L)

81 MAT   P =  B*V

100 DATA 10,5
110 DATA .25, .25, 0,0,0
111 DATA .25,0,0,0,0
112 DATA 0,0,0,0,0
113 DATA 0,.25,0,0,0
114 DATA 0,0,0,0,0
115 DATA 0,0,.25,0,0
116 DATA 0,0,0,.25,0
```

```
117 DATA 0,0,0,0,0
118 DATA 0,0,.25,0,.25
119 DATA 0,0,0,.25,.25
120 DATA 0,.25,0,.25,0,0,0,0,0,0
121 DATA .25,0,.25,0,.25,0,0,0,0,0
122 DATA 0,.5,0,0,0,.5,0,0,0,0
123 DATA .25,0,0,0,.25,0,.25,0,0,0
124 DATA 0,.25,0,.25,0,.25,0,.25,0,0
125 DATA 0,0,.25,0,.25,0,0,0,.25,0
126 DATA 0,0,0,.25,0,0,0,.25,0,.25
127 DATA 0,0,0,0,.25,0,.25,0,.25,25
128 DATA 0,0,0,0,0,.25,0,.25,0,0
129 DATA 0,0,0,0,0,0,.25,.25,0,0
130 DATA +5,–5,–5,–5,+10

READY

RUN
```

TRANS3:

```
–0.17    +1.08    –0.02    –1.78    –0.46
–1.13    –1.48    –0.03    +0.96    +0.87

0.292 SEC.
READY
```

We find that in state 5 one almost but not quite breaks even. There is an overall expected loss of 46 cents; thus one should not agree to play the game starting at state 5.

Although we gave this application an interpretation as a game, there are many other interpretations of a Markov chain game. There are processes in nature that are described by absorbing Markov chains where one can in a natural way assign a "value" to ending up at a given terminal. Here the expected pay-off has a natural interpretation. There is also a method for computing voltages in a simple electric circuit using this technique. In recent years Markov chains have acquired considerable importance in applications to many sciences. It is, therefore, interesting to see how easy it is to compute fundamental quantities for Markov chains by means of a high-speed computer.

Exercises

1. Vectors in BASIC are column vectors, but a matrix of one row may be used as a row vector. Write a programme to read a row vector and a column vector, of four components each, and to compute their product.
2. If A is probability row vector, and if it is repeatedly multiplied on the right by a regular transition matrix P, it will approach the fixed vector. Write a programme to carry out this process.
3. For a transient chain, TV may be computed as the sum of the infinite series

$$N = I+Q + Q^2 + Q^3 + ...$$

   Write a programme to compute the first 21 terms of this series.
4. Write a programme which, for an absorbing Markov chain, will compute Q and R from N and B. *[Hint:* Compute N^{-1}.]
5. Compute powers of the transition matrix

$$P = \begin{pmatrix} .1 & .2 & .3 & .4 \\ .4 & .3 & .2 & .1 \\ .3 & .1 & .4 & .2 \\ .2 & .4 & .1 & .3 \end{pmatrix}$$

   to find the fixed vector.
6. Compute powers of the transition matrix

$$P = \begin{pmatrix} 0 & .3 & 0 & .7 \\ .5 & 0 & .5 & 0 \\ 0 & .6 & 0 & .4 \\ .2 & 0 & .8 & 0 \end{pmatrix}$$

   and explain the result.
7. Let h be an arbitrary column vector of three components. Multiply it repeatedly by the OZ transition matrix and observe that it tends to a constant vector. Interpret the constant.
8. Try out the programme of Exercise 2 for the Land of Oz.

9. RUN the programme of Exercise 4 to verify that it produces Q and R correctly.
10. Modify the programme TRANS to verify the identity $QB = B - R$.
11. Design your own Markov chain game and compute the expected values for various starting positions.

Linear Equations

The purpose of this section is to translate the flow diagram for the solution of linear equations into a computer programme. We shall first do this in a straightforward manner and show that the programme reproduces the results. However, we shall then note that the programme is inadequate; this will give us an opportunity to consider one of the deeper problems of computer programming— namely, the question of numerical accuracy. We shall assume that there is no variable all of whose coefficients are 0.

The programme, LINEQU is designed to follow Fig 2.5 of the previous chapter. Boxes 1-6 correspond to blocks of instructions starting at lines 100, 200, 300, 400, 500, and 600. Box 7 is combined with box 2, and box 8 with box 5. For easy identification each block of instructions starts with a REM (or "remark") statement. Such a REM statement is for the convenience of the programmer and is ignored by the computer. LINEQU is designed to solve M equations in N unknowns. We start by saving space for our list and table, reading M and N, and then reading the tableau T of coefficients. It should be noted that T has N + 1 columns since it contains not only the coefficients of the left-hand side of the equations but also the numbers on the right-hand side.

In the remainder of the programme I and J will be the subscripts corresponding to the pivot. The auxiliary variables I1 and J1 are used as running subscripts for rows and columns. The process consists of choosing a pivot in each row and operating with it; this loop starts at line 100 and ends at line 500. Lines 200-220 search for a non-zero element in row I. If such an element is found, we jump to line 300. It should be noted that as we jump out of the loop of lines 200-220, the subscript J is correctly set for the pivot. If we complete the entire loop, then the left-hand side of the equation is zero. In line 230 we check whether the right-hand side is also zero. If it is not, we jump to line 900 and type out "THERE IS NO SOLUTION."

At line 300 we note what the pivot is. We also put into the list P the subscript of the variable we pivoted on. This will make it much easier to identify the solution of the problem. It should be noted that if the equation was identically equal to zero, and hence could be ignored, then at line 240 we entered a zero into the list of pivots.

LINEQU:

```
5 DIM T(20, 21), P(20)
10 READ M, N
20 MAT READ T(M, N+1)
29
100 REM START MAIN LOOP
110 FOR I = 1 TO M
120
200 REM  FIND PIVOT
205 FOR J = 1 TO N
210 IF T(I, J)< > 0 THEN 300
220 NEXT J
230 IF T(I, N+1) < > 0 THEN 900
240 LET P(I) = 0
250 GOTO 500
260
300 REM DIVIDE BY PIVOT
302 LET P = T(I, J)
305 LET P(I) = J
310 FOR J1 = 1 TO N+1
320 LET T(I, J1) = T (I, J1)/P
330 NEXT J1
340
400 REM SUBTRACT MULTIPLES OF ROW
405 FOR I1 = 1 TO M
410 IF I1 = I THEN 460
420 LET C = T(I1, J)
430 FOR J1 = 1 TO N+1
440 LET T(I1, J1) = T(I1, J1) – C*T(I, J1)
450 NEXT J1
460 NEXT I1
470
```

```
500 CLOSE MAIN LOOP
510 NEXT I
520
600 REM PRINT ANSWER
605 FOR I = 1 TO M
610 LET P = P(I)
620 IF P = 0 THEN 790
625 LET B = T(I, N+1)
630 PRINT "X"; STR$(P); "="; STR$(B);
640 FOR J = 1 TO N
645 IF J = P THEN 690
650 LET C = T(I, J)
660 IF C = 0 THEN 690
665 IF C<0 THEN 680
670 PRINT "–";
675 GOTO 687
680 LET C = –C
685 PRINT "+";
687 PRINT STR$(C); "*"; "X"; STR$(J);
690 NEXT J
700 PRINT
790 NEXT I
800 GOTO 999
900 PRINT "THERE IS NO SOLUTION."
905
910 DATA 3,3
920 DATA 1,4,3,1
921 DATA 2,5,4,4
922 DATA 1,–3,–2,5
999 END
READY

RUN
```

LINEQU:

```
XI = 3.
X2 = –2.
X3 = 2.

0.166 SEC.
READY
```

Lines 310-330 complete this particular box of the flow diagram by dividing all coefficients in this row of the tableau by the pivot P.

We must now subtract suitable multiples of the pivotal row from the other rows. This is accomplished in lines 400-460. The subscript I1 will run through all the rows; however, on line 410 we make sure that we skip over the pivotal row. C is equated to the appropriate multiplier and the subtraction is carried out in the loop in lines 430-450. When this double loop is completed, on line 500 we go on to the next row. It should be noted that the entire heart of the programme is contained in lines 100-500, a total of only 20 instructions in BASIC.

The answers are printed in the loop of lines 600-790. This piece of code could be much simpler except for two complications: First, we want to have a nice format for the answer. Second, we want to handle not only the case of a unique solution but find all possible solutions in case there are infinitely many of them. To obtain a nice-looking form for the answers, we need to introduce an additional feature of BASIC (It should be noted that there are many other advanced features of BASIC not covered in this book). When BASIC prints the numerical value of a variable, it either starts with a minus sign or a blank and places a blank after the number. This is very convenient when we simply want to print a list of numbers one after the other. However, it spoils the output when we want to print, for example, "X5." But writing the string command "STR$(P)" will print a numerical value of P without initial or trailing blanks (This command is not available in all versions of BASIC).

If the solution were always unique, the output would be accomplished by the six instructions in lines 600-630 and 790. We look at each equation once, and look up in the list of pivots what the subscript P of the pivot was. If this is 0, then the equation is identically 0 and therefore can be ignored. Otherwise B is the value of the variable and, on line 630, we print out an answer that may look like "X5 = 3.2." Here is where we see the advantage of having remembered the element we pivoted on. Its coefficient ends up being 1, and hence the right-hand side of the equation is its value.

The loop in lines 640-690 handles the case of infinitely many solutions. If the solution is not unique, then variables other than the pivot are left over with non-zero coefficients. These variables may be given arbitrary values. We usually indicate this by "bringing the variable to the right-hand side." Thus, we search in the loop to see whether any variable other than the pivot has a non-zero coefficient. If it does, we print it with a suitable coefficient after the value we have already printed on the right-hand side. It should be noted that we had to test on line 665 whether the coefficient was positive or negative and the two cases are treated separately. It is left as an exercise for the reader to step through this part of the programme by hand.

```
920 DATA 1,–2,–3, 2
921 DATA 1,–4,–13,14
922 DATA –3,5,4,0
READY

RUN
```

LINEQU2:

```
XI = –10 - 7*X3
X2 = –6 – 5*X3

0.153 SEC.
READY
```

To show that the programme also works in the case of infinitely many solutions or no solutions, we change the data to those of Examples 2 and 3 respectively. These are shown in LINEQU2 and LINEQU3. It should be noted that the output format for the case of infinitely many solutions is very easily readable. It shows that X3 may take on any value and it indicates what the corresponding values of X1 and X2 must be.

```
922 DATA –3,5,4,2

READY

RUN
```

LINEQU3:

```
THERE IS NO SOLUTION.

0.167 SEC.
READY
```

Next we try out the programme with a larger data base. In LINEQU4, we have four equations in five unknowns.

```
910 DATA 4, 5
920 DATA 3, 2, 1, 2, 3, 11
921 DATA –l, –l, –2, l, l, –6
922 DATA 1, 2, 3, 4, 5, 3
923 DATA 2, 2, 0, 8, 10, 2
READY
```

RUN

LINEQU4:

```
XI = 4.5 + 3.5*X4

X2 = -1. – 7.5*X4

X3 = 1 + 2.5*X4

X5 = –0.5

0.179 SEC.
READY
```

The result seems reasonable; indeed, if we check it all the indicated solutions are correct. However, a more careful check will show that we have failed to find all the solutions! We have run into one of the subtleties of computer programming: a programme that to all appearances is correct produces incorrect results. The problem is one of round-off errors. When this happens, one must do troubleshooting on the computer programme, or as it is commonly phrased, one must "debug" it. A very useful procedure is to ask the computer to print out not just the final solution but also the intermediate steps. This may be accomplished by replacing line 500 by:

```
500 MAT PRINT T;
```

With this change the tableau will be printed after each iteration. A look at the output would indicate that something went wrong between the third and fourth iterations. The last line in the third iteration would appear as follows:

$$2 \times 10^{-7} \times X5 = -1 \times 10^{-7}.$$

From this the computer concludes that X5 equals —0.5. However, the very small numbers appearing in this equation represent round-off errors and should actually be zero. The reason for round-off errors is the fact that a computer can only carry a fraction to a limited number of decimal places (usually six to nine). Furthermore, it works with a number system to base 2. Whether a rounding is necessary depends on the base. Therefore one must anticipate that in hand calculations where no round-off error appears, one may appear on the computer, or vice versa. In this particular case this very minute round-off error changes the whole nature of the solution. The equation should actually be 0 = 0, and therefore X5 should be available as a variable whose value may be chosen arbitrarily. Therefore, due to a minute error, we lost infinitely many available solutions.

The lesson that we learn is that if a variable's value was computed through complicated calculation, we cannot assume that a 0 will come out to be exactly 0. This forces us to modify lines 210, 230, and 660. We shall assume that any sufficiently small number is produced by a round-off error and should really be a 0. Of course, the question is just what does "sufficiently small" mean? This is a deep and difficult question, and there is no universally satisfactory answer to it. For any proposed solution to this problem one can find a set of equations for which the programme will produce the wrong results. We shall, however, show one quite common solution to this dilemma that will handle "normal" cases. Our assumption will be that any coefficient that turns out to be less than 10^{-6} is a round-off error and should be 0.

Thus, on line 660 instead of asking whether the coefficient C is equal to 0, we shall ask whether its absolute value is less than 10^{-6}. In BASIC one computes the absolute value of C by writing "ABS(C)." The corresponding corrections must also be made on line 210 and line 230. We show these corrections and the corrected run in LINEQU5. This time we have found all the solutions to the problem.

```
210 IF ABS(T(I, J)) > 1E–6 THEN 300
230 IF ABS(T(I,N+1)) > 1E-6 THEN 900
660 IF ABS(C) < 1E–6 THEN 690
READY

RUN
```

LINEQU5:

```
XI = 6.5 + 3.5*X4 + 4.*X5
X2 = –5.5 – 7.5*X4 – 9.*X5
X3 = 2.5 + 2.5*X4 + 3.*X5

0.202 SEC.
READY
```

The resulting programme is of quite general use in solving linear equations. The exercises will show modifications of this programme that may be used for other purposes—e.g., inverting a matrix. We again see that a relatively short BASIC programme, and surprisingly short computing times, can solve important practical problems.

This section has also given the reader a first taste of the complex field of finding numerical solutions to mathematical problems. This field is known as *numerical analysis.* It treats the wide variety of difficulties one runs into in finding numerical solutions, and also searches for the most efficient numerical methods of solving a variety of problems.

1. Modify the programme LINEQU5 to solve simultaneously two sets of equations with identical left sides but different right sides.
2. If the coefficients of the left side of a set of equations form an $n \times n$ matrix A, and if one successfully pivots on every row (no left side becomes identically zero), then A^{-1} exists. This is true irrespective of the right side of the equation. Modify LINEQU5 to serve as a test of whether a given square matrix has an inverse.
3. Use LINEQU5 to solve the following equations:

$$4X_1 - 3X_2 + 2X_3 - X_4 + 5X_5 = -10$$
$$X_1 - 2X_2 + 3X_3 + X_4 - X_5 = 12$$
$$2X_1 + X_2 - X_3 + 2X_4 + 5X_5 = -1$$
$$3X_1 - 2X_2 + X_3 - X_4 + 2X_5 = -6$$

[Ans. 1,2,3,4,–2.]

4. RUN the programme of Exercise 1 using the DATA of LINEQU2 and LINEQU3.

Statistics

Introduction

In the study of probability theory, we assign a probability measure to the possible outcomes of an experiment. We then make probability predictions relating to the experiment. For example, a coin is tossed ten times. We assign an equal weight to all possible sequences of heads and tails. We then compute the probability that exactly six heads turn up. We find that this probability is $\binom{10}{6} \cdot \left(\frac{1}{2}\right)^{10} = .205$. Statistics deals with the inverse problem.

We do not know the basic probability measure, but we are able to carry out certain chance experiments, from which we obtain information about the underlying measure.

As an example, assume that in a large population each person holds an opinion on the question of legalising marijuana. They either favour this or are opposed. We choose at random 20 people and ask them their opinions. Choosing "at random" means that we have an equal chance of obtaining any group of 20 people from the entire population. If the size of the population is large, the effect of knowing certain of the opinions will not significantly change the chance that the next person sampled will say "yes." Thus, it is reasonable to assume that the underlying chance model is an independent trials model with probability p for success (answer "yes") on each trial, where p is the proportion in the entire population that favour legalising marijuana.

On the basis of the sample, we would like to estimate p. The intuitive estimate for the parameter p would be simply the fraction $\overline{p}$ of persons in the sample that say "yes." In Figure 4.1 we show the result of drawing ten samples of 20 each in a case where $p = .4$. While our estimates are in general near the true value .4, our worst estimate is .2, only half the true value.

Experiment number	*Number of "yes" answers*	*Fraction* $\overline{p}$
1	6	.3
2	8	.4
3	9	.45
4	4	.2
5	7	.35
6	6	.3
7	6	.3
8	9	.45
9	5	.25
10	7	.35

From the binomial measure, we can calculate the exact probability that the observed fraction $\overline{p}$ will lie in a given range. For example, the values of $f(20, x; .4)$ for x between 6 and 10 add up to .747. Thus, with probability .747, our estimate will be between .3 and .5.

We recall from our study of independent trials that the expected number of successes in a sample of size n is np and the standard deviation for the number of successes is $\sqrt{npq}$. Further, the probability of a deviation of more than 3 standard deviations from the expected number is very unlikely (.001). Thus, if we increase the sample size to 2400 the expected number of "yes" responses would be 960 and the standard deviation $\sqrt{2400 \times .4 \times .6} = 24$. Thus, our estimate would with high probability lie between $\frac{960-72}{2400}=.37$ and $\frac{960+72}{2400}=.43$, in the interval [.37, .43].

This suggests that when p is unknown we should try to estimate from the sample an interval within which we believe the true p lies.

In some situations, we need to make a choice between two estimates for p. For example, the incidence of colds may be known and we wish to test the claim that this can be decreased if people take large doses of vitamin C. Thus, we have to determine whether the incidence of colds among those taking vitamin C is the same as for the whole population or a smaller value. Or a manufacturer may assume that his production process is operating correctly if it produces no more than 1 per cent defective items but is not operating correctly if it produces as many as 5 per cent defective items. He is interested in devising a test to see if the system is operating correctly.

Perhaps the largest statistical test ever conducted was the test designed in the early '50s to see if the vaccine developed by Jonas Salk would effectively cut down the incidence of polio. The average incidence of polio at that time was about 50 per 100,000 persons. It was not expected that the vaccine would be 100 per cent effective, but it was hoped that it would cut down the incidence of polio by at least 50 per cent. Thus, we can view the experiment as a test of the hypothesis that a person vaccinated will have a significantly lower probability of being afflicted with polio than a person not vaccinated.

In applying probability models, predictions from the model are only reliable if the assumptions made in describing the model are reasonably met. Similarly, our statistical inferences are based upon certain assumptions. We have already mentioned the assumption of randomness in a sample. There are many pitfalls that one can fall into if care is not taken. Perhaps the most famous example of this is the celebrated prediction of the *Literary Digest* that Alfred Landon would defeat Franklin Roosevelt in the 1936 presidential election. In this poll, the sample was chosen from names obtained from telephone books and car registrations. In 1936, this was not at all a "random sample" and the prediction was badly in error. Opinion polls are still trying to recover from this blunder.

Before we continue our discussion of statistics we shall need one important result from probability theory called the central limit theorem.

For use in the exercises and in later sections, we show the probabilities for ten independent trials and various values of f in Figure 4.2.

Table of values of $f(10, x; p)$

X	0.1	0.25	0.4	0.5	0.6	0.75	0.9
0	0.349	0.056	0.006	0.001	0	0	0
1	0.387	0.188	0.040	0.010	0.002	0	0
2	0.194	0.282	0.121	0.044	0.011	0	0
3	0.057	0.250	0.215	0.117	0.042	0.003	0
4	0.011	0.146	0.251	0.205	0.111	0.016	0
5	0.001	0.058	0.201	0.246	0.201	0.058	0.001
6	0	0.016	0.111	0.205	0.251	0.146	0.011
7	0	0.003	0.042	0.117	0.215	0.250	0.057
8	0	0	0.011	0.044	0.121	0.282	0.194
9	0	0	0.002	0.010	0.040	0.188	0.387
10	0	0	0	0.001	0.006	0.056	0.349

Exercises

1. A random sample of ten persons is chosen in New York City at a time when 60 per cent are in favour of Kelly for mayor and 40 per cent are in favour of McGrath. What is the probability that the sample will show less than 50 per cent in favour of Kelly?

   [*Ans.* .166]

2. An independent trials experiment is repeated ten times with six successes. Which value of p in Figure 4.2 gives the highest probability of obtaining the outcome of six successes—i.e., the observed outcome?

3. In Exercise 1, assume that the sample size is increased to 9600. Find the expected number and the standard deviation for the number of those in favour of Kelly. What could we say about the range

of our estimates if the number of "yes" responses does not deviate by more than three standard deviations from the expected number?

4. In a city, there are 100,000 persons who are going to vote on the question of legalising marijuana. Of these, 90,000 are under 50 years of age and 10,000 are 50 or over. Assume that 75 per cent of those under 50 favour legalising marijuana and of those 50 or older only 20 per cent are in favour. What is the probability that a person chosen at random will favour legalising marijuana? In a sample of 100 chosen at random what is the expected number that will answer "yes"? What is the expected number if a random sample of 100 is chosen, 50 from each of the two groups?

5. In an experiment where the probability distribution depends on a single number, or parameter, the following is a standard method of estimating this parameter. Choose the value of the parameter which gives the highest probability of obtaining the observed result. This method is called the method of *maximum likelihood.* On the basis of the result of Exercise 2, what would you guess to be the maximum likelihood estimator for an independent trials experiment for the probability p of success when x successes are observed in n trials?

6. A box has ten items, eight good and two defective. A sample of five is chosen with replacement—that is, after each item is chosen and inspected it is replaced (i.e., put back) before the next item is drawn. Find the probability that the sample has exactly one defective item.

[*Ans.* .410]

7. Answer the same question as in Exercise 6 if the sampling is done without replacement. That is, a set of five is chosen at random from all possible subsets of five items of the box. Find the probability that the sample has exactly one defective item and compare your answer to that obtained in Exercise 6.

[Ans. $\dfrac{\binom{2}{1}\binom{8}{4}}{\binom{10}{5}}$ =.556]

8. A sample of three items is chosen from a box of 1000 of which 80 per cent are defective. Show that the probability of obtaining

exactly one defective item is essentially the same whether we sample with or without replacement.

9. Assume that the incidence of lung cancer among smokers is estimated to be 20 per 100,000 and among heavy smokers to be 200 per 100,000. Estimate the probability that a person who smokes will *not* get lung cancer and compare this with the estimate for a heavy smoker.

[*Ans.* .9998, .998]

10. A hardware store receives boxes of 50 bolts. Experience has shown that they occasionally get a bad lot. When they get a box they choose two bolts at random and if either is defective they return the box. Assume that a box has five defective bolts. What is the probability that the box will be sent back?

11. Referring to Exercise 10, assume that the store receives shipments of ten boxes each with 50 bolts. It combines the 500 bolts and then chooses two bolts at random; if either is defective it sends back the entire lot. If the shipment contains 50 defectives in all, is the probability of the lot being returned larger than, equal to, or smaller than the probability in Exercise 10 of a single box with five defectives being returned?

12. Toss a coin 100 times. In each group of ten tosses, count the number of heads. Compare the results with Figure 4.2.

13. Toss a pair of coins 100 times. In each group of ten tosses, count the number of times two heads turn up. Compare the results with Figure 4.2.

Central Limit Theorem

As we have indicated, to go further in our discussion of statistics, we shall need an important theorem of probability theory called the *central limit theorem.* While this is a very general theorem, we shall discuss it in this section only as it applies to independent trials processes.

As usual, let p be the probability of success on a trial *and* $f(n, p; x)$ the probability of exactly x successes in n trials.

In Figure 4.3 we have plotted bar graphs which represent $f(n, 3; x)$ for $n = 10, 50, 100$, and 200. We note first of all that the graphs are drifting off to the right. This is not surprising, since their peaks occur at np, which is steadily increasing. We also note that while the total area is always 1, this area becomes more and more spread out.

We want to redraw these graphs in a manner that prevents the drifting and the spreading out. First of all, we replace x by $x - np$, assuring that our peak always occurs at 0. Next, we introduce a new unit for measuring the deviation, which depends on n, and which gives comparable scales. As we saw earlier that the standard deviation $\sqrt{npq}$ is such a unit.

We must still insure that probabilities are represented by areas in the graph. In Figure 4.3, this is achieved by having a unit base for each rectangle, and having the probability $f(n,p; x)$ as height. Since we are now representing a standard deviation as a single unit on the horizontal axis, we must take $f(n, p; x)\sqrt{npq}$ as the heights of our rectangles. The resulting curves for $n = 50$ and $n = 200$ are shown in Figures 4.4 and 4.5, respectively.

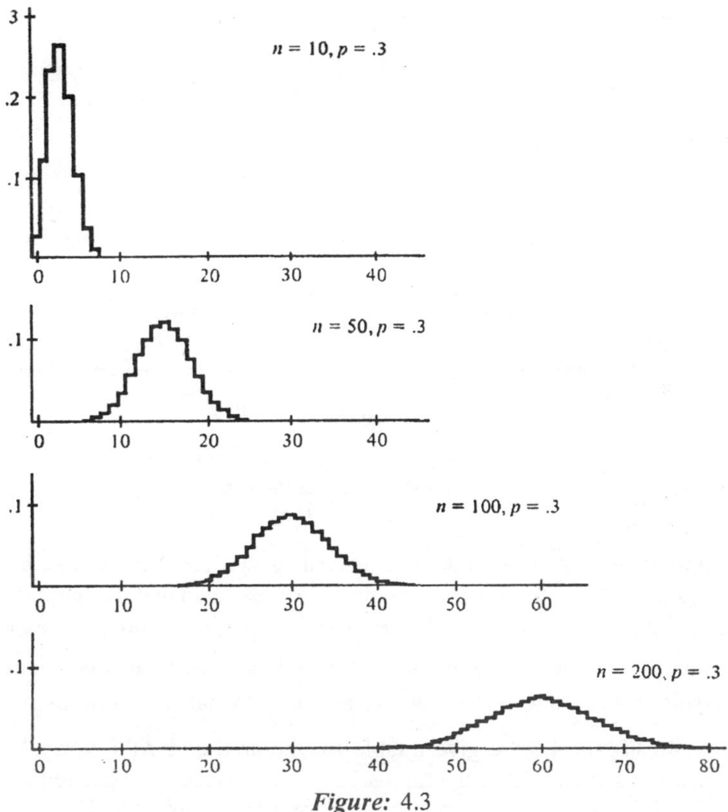

Figure: 4.3

We note that the two figures look very much alike. We have also shown in Figure 4.5 that it can be approximated by a bell-shaped curve. This curve represents the function

$$f(x) = \frac{1}{\sqrt{2\pi}} e^{-x^2/2},$$

and is known as the *normal curve.* It is a fundamental theorem of probability theory that as n increases, the appropriately rescaled bargraphs more and more closely approach the normal curve. The theorem is known as the *central limit theorem,* and we have illustrated it graphically.

More precisely, the theorem states that for any two numbers a and b, with $a < b$,

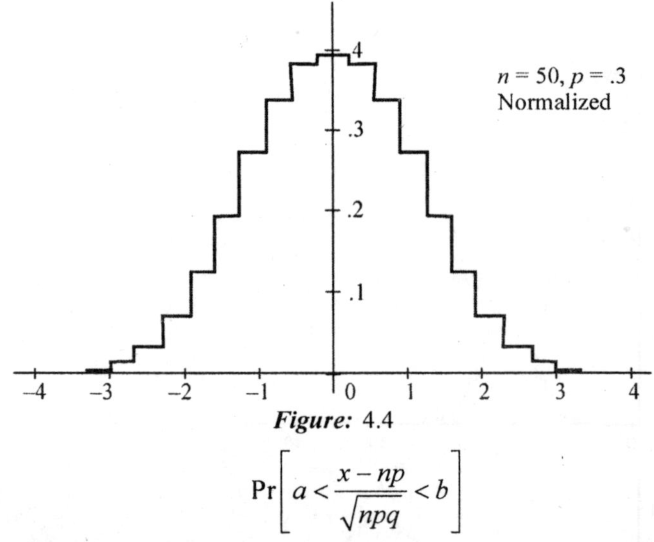

Figure: 4.4

$$\Pr\left[a < \frac{x - np}{\sqrt{npq}} < b\right]$$

approaches the area under the normal curve between a and b, as n increases. This theorem is particularly interesting in that the normal curve is symmetric about 0, while $f(n, p|x)$ is symmetric about the expected value np only for the case $p = \frac{1}{2}$. It should also be noted that we always arrive at the same normal curve, no matter what the value of p is.

In Figure 4.6 we give a table for the area under the normal curve between 0 and d. Since the total area is 1, and since it is symmetric about the origin, we can compute arbitrary areas from this table. For example, suppose that

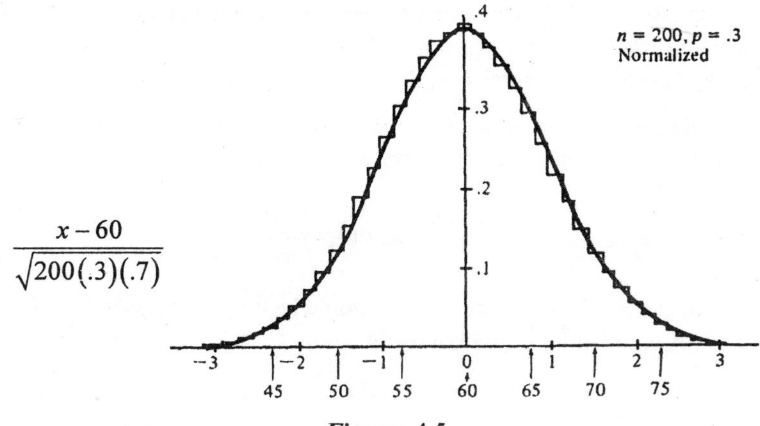

Figure: 4.5

$A(d)$ = area of shaded region

0 d

d	*A(d)*	*d*	*A(d)*	*d*	*A(d)*	*d*	*A(d)*
.0	.000	1.1	.364	2.1	.482	3.1	.4990
.1	.040	1.2	.385	2.2	.486	3.2	.4993
.2	.079	1.3	.403	2.3	.489	3.3	.4995
.3	.118	1.4	.419	2.4	.492	3.4	.4997
.4	.155	1.5	.433	2.5	.494	3.5	.4998
.5	.191	1.6	.445	2.6	.495	3.6	.4998
.6	.226	1.7	.455	*in*	.497	3.7	.4999
.7	:258	1.8	.464	2.8	.497	3.8	.49993
.8	.288	1.9	.471	2.9	.498	3.9	.49995
.9	.316	2.0	.477	3.0	.4987	4.0	.49997
1.0	.341					5.0	.49999997

Figure: 4.6

we wish the area between -1 and $+2$. The area between 0 and 2 is given in the table as .477. The area between -1 and 0 is the same as between 0 and 1, and hence is given as .341. Thus, the total area is .818. The area outside the interval $(-1, 2)$ is then $1 - .818 = .182$.

Example 1: Let us find the probability that s differs from the expected value np by as much as d standard deviations.

$$\Pr\left[|x-np| \geq d\sqrt{npq}\right] = \Pr\left[\left|\frac{x-np}{\sqrt{npq}}\right| \geq d\right],$$

and hence the approximate answer should be the area outside the interval $(-d, d)$ under the normal curve. For $d = 1, 2, 3$, we obtain

$$1 - (2 \times .341) = .318, \quad 1 - (2 \times .477) = .046$$

and

$$1 - (2 \times .4987) = .0026,$$

respectively.

Example 2: We considered the example of tossing a coin 10,000 times. The expected number of heads that turn up is 5000, and the standard deviation is $\sqrt{10{,}000 \cdot \frac{1}{2} \cdot \frac{1}{2}} = 50$. We observed that the probability of a deviation of more than two standard deviations (or 100) is very unlikely.

On the other hand, consider the probability of a deviation of less than .1 standard deviation—that is, of a deviation of less than 5. The area from 0 to .1 under the normal curve is .040, and hence the probability of a deviation from 5000 of less than 5 is approximately .08. Thus, while a deviation of 100 is very unlikely, it is also very unlikely that a deviation of less than 5 will occur.

Example 3: The normal approximation can be used to estimate the individual probability $f(n, x; p)$ for large n. For example, let us estimate $f(200, 65; .3)$. The graph of the probabilities $f(200, x; .3)$ was given in Figure 4.5 together with the normal approximation. The desired probability is the area of the bar corresponding to $x = 65$. An inspection of the graph suggests that we should take the area under the normal curve between

64.5 and 65.5 as an estimate for this probability. In normalised units, this is the area between

$$\frac{4.5}{\sqrt{200(.3)(.7)}} \text{ and } \frac{5.5}{\sqrt{200(.3)(.7)}},$$

or between .6944 and .8487. Our table is not fine enough to find this area, but from more complete tables, or by machine computation, this area may be found to be .046 to three decimal places. The exact value to three decimal places is .045. This procedure gives us a good estimate.

If we check all of the values of $f(200, x; .3)$ we find in each case that we would make an error of at most .001 by using the normal approximation. There is unfortunately no simple way to estimate the error caused by the use of the central limit theorem. The error will clearly depend upon how large n is, but it also depends upon how near p is to 0 or 1. The greatest accuracy occurs when p is near $\frac{1}{2}$.

Exercises

1. Let x be the number of successes in n trials of an independent trials process with probability p for success. Let $x^* = \dfrac{x - np}{\sqrt{npq}}$. For large n estimate the following probabilities:
   (a) $\Pr[x^* < -2.5]$. [*Ans.* .006]
   (b) $\Pr[x^* < 2.5]$.
   (c) $\Pr[x^* \geq -.5]$.
   (d) $\Pr[-1.5 < x^* < 1]$. [*Ans.* .774]
2. A coin is biased in such a way that a head comes up with probability .8 on a single toss. Use the normal approximation to estimate the probability that in a million tosses there are more than 800,400 heads.
3. Plot a graph of the probabilities $f(10, x; .5)$. Plot a graph also of the normalised probabilities as in Figures 4.4 and 4.5.
4. An ordinary coin is tossed 1 million times. Let x be the number of heads which turn up. Estimate the following probabilities:
   (a) Pr $[499{,}500 \leq x \leq 500{,}500]$.
   (b) Pr $[499{,}000 \leq x \leq 501{,}000]$.

(c) $\Pr[498{,}500 \le x \le 501{,}500]$.

[*Ans.* .682; .954; .997 (approximate answers)]

5. Assume that a baseball player has probability .37 of getting a hit each time he comes to bat. Find the probability of getting an average of .388 or better if he comes to bat 300 times during the season. (In 1957, Ted Williams had a batting average of .388 and Mickey Mantle had an average of .353. If we assume this difference is due to chance, we may estimate the probability of a hit as the combined average, which is about .37)

[*Ans.* .242]

6. A true-false examination has 48 questions. Assume that the probability that a given student knows the answer to any one question is $\frac{3}{4}$. A passing score is 30 or better. Estimate the probability that the student will fail the exam.

7. We assume that the school decides to admit 1296 students. Estimate the probability that they will have to have additional dormitory space.

[*Ans.* Approximately .115]

8. Peter and Paul each have 20 pennies. They each toss a coin and Peter wins a penny if his coin matches Paul's, otherwise he loses a penny; they do this 400 times, keeping score but not paying until the 400 matches are over. What is the probability that one of the players will not be able to pay? Answer the same question for the case in which Peter has 10 pennies and Paul has 30.

9. In tossing a coin 100 times, the probability of getting 50 heads is, to three decimal places, .080. Estimate this same probability using the central limit theorem. [*Ans.* .080]

10. A standard medicine has been found to be effective in 80 per cent of the cases where it is used. A new medicine for the same purpose is found to be effective in 90 of the first 100 patients on which the medicine is used. Could this be taken as good evidence that the new medication is better than the old?

11. Two railroads are competing for the passenger traffic of 1000 passengers by operating similar trains at the same hour. If a given passenger is equally likely to choose one train as the other, how many

seats should the railroad provide if it wants to be sure that its seating capacity is sufficient in 99 out of 100 cases? [*Ans.* 537]

Test of Hypotheses

We turn now to our first typical statistical problem. As we indicated in the introductory section, our problem is often to decide between two or more competing probability measures. We shall illustrate this in terms of an example.

Example: Smith claims that he has the ability to distinguish ale from beer and has bet Jones a dollar to that effect. Now Smith does not mean that he can distinguish beer from ale with 100 per cent accuracy, but rather that he believes that he can distinguish them a proportion of the time which is significantly greater than $\frac{1}{2}$.

Assume that it is possible to assign a number p which represents the probability that Smith can pick out the ale from a pair of glasses, one containing ale and one beer. We identify $p = \frac{1}{2}$ with his having no ability, $p > \frac{1}{2}$ with his having some ability, and $p < \frac{1}{2}$ with his being able to distinguish, but having the wrong idea which is the ale. If we knew the value of p, we would award the dollar to Jones if p were $\leq \frac{1}{2}$, and to Smith if p were $> \frac{1}{2}$. As it stands, we have no knowledge of p and thus cannot make a decision. We perform an experiment and make a decision as follows.

Smith is given a pair of glasses, one containing ale and the other beer, and is asked to identify which is the ale. This procedure is repeated ten times, and the number of correct identifications is noted. If the number correct is at least eight, we award the dollar to Smith, and if it is less than eight, we award the dollar to Jones.

We now have a definite procedure and shall examine this procedure from both Jones's and Smith's points of view. We can make two kinds of errors. We may award the dollar to Smith when in fact the appropriate value of p is $< \frac{1}{2}$, or we may award the dollar to Jones when the appropriate value for p is $> \frac{1}{2}$. There is no way that these errors can be completely avoided. We hope that our procedure is such that each of the bettors will be convinced that, if he is right, he will very likely win the bet.

Jones believes that the true value of p is $\frac{1}{2}$. We shall calculate the probability of Jones winning the bet if this is indeed true. We assume that the individual tests are independent of each other and all have the same probability $\frac{1}{2}$ for success. (This assumption will be unreasonable if the glasses are too large.) We have then an independent trials process with $p = \frac{1}{2}$ to describe the entire experiment. The probability that Jones will win the bet is the probability that Smith gets fewer than eight correct. From the table in Figure 4.2 we compute that this probability is approximately .945. Thus, Jones sees that, if he is right, it is very likely that he will win the bet.

Smith, on the other hand, believes that p is significantly greater than $\frac{1}{2}$. If he believes that p is as high as .9, we see from Figure 4.2 that the probability of his getting eight or more correct is .930. Then, both men will be satisfied by the bet.

Suppose, however, that Smith thinks the value of p is only about .75. Then the probability that he will get eight or more correct and thus win the bet is .526. There is then only an approximately even chance that the experiment will discover his abilities, and he probably will not be satisfied with this. If Smith really thinks his ability is represented by a p value of about $\frac{3}{4}$, we would have to devise a different method of awarding the dollar. We might, for example, propose that Smith win the bet if he gets seven or more correct. Then, if he has probability $\frac{3}{4}$ of being correct on a single trial, the probability that he will win the bet is approximately .776. If $p = \frac{1}{2}$, the probability that Jones will win the bet is about .828 under this new arrangement. Jones's chances of winning are thus decreased, but Smith may be able to convince him that it is a fairer arrangement than the first procedure.

In the theory of hypothesis testing it is common to refer to one hypothesis, say $p = \frac{1}{2}$, as the null hypothesis H_0, and an alternate hypothesis as H_1.

In the above example, it was possible to make two kinds of errors. The probability of making these errors depended on the way we designed the experiment and the method we used for the required decision. In some cases, we are not too worried about the errors and can make a relatively simple experiment. In other cases, errors are very important, and the experiment must be designed with that fact in mind. For example, the possibility of error is certainly important in the case that a vaccine for a given disease is proposed and the statistician is asked to help in deciding whether or not it should be used. In this case, it might be assumed that there is a certain probability p that a person will get the disease if not vaccinated and a probability r that he will get it if he is vaccinated. If we have some knowledge of the approximate value of p, we are then led to construct an experiment to decide whether r is greater than p, equal to p, or less than p. The first case would be interpreted to mean that the vaccine actually tends to produce the disease, the second that it has no effect, and the third that it prevents the disease; so that we can make three kinds of errors. We could recommend acceptance when it is actually harmful, we could recommend acceptance when it has no effect, or finally we could reject it when it actually is effective. The first and third might result in the loss of lives, the second in the loss of time and money of those administering the test. Here it would certainly be important that the probability of the first and third kinds of errors be made small. To see how it is possible to make the probability of both errors small, we return to the case of Smith and Jones.

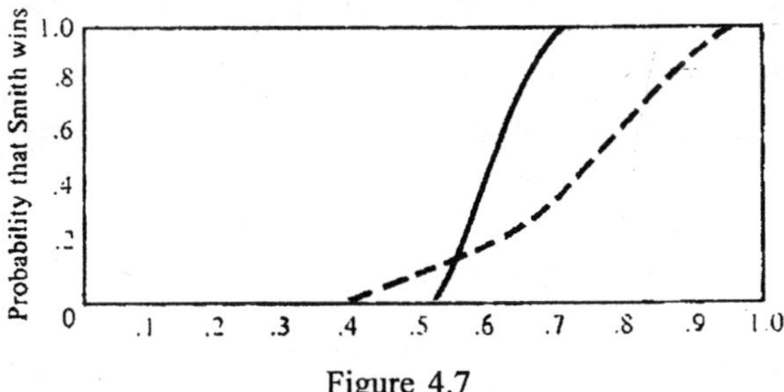

Figure 4.7

Example: Suppose that, instead of demanding that Smith make at least eight correct (continued) identifications out of ten trials, we insist that he make at least 60 correct identifications out of 100 trials. (The glasses must now be very small.) Then, if $p = \frac{1}{2}$, the probability that Jones wins the bet is about .98; so that we are extremely unlikely to give the dollar to Smith when in fact it should go to Jones. (If $p < \frac{1}{2}$, it is even more likely that Jones will win.) If $p > \frac{1}{2}$, we can also calculate the probability that Smith will win the bet. These probabilities are shown in the graph in Figure 4.7. The dashed curve gives for comparison the corresponding probabilities for the test requiring eight out of ten correct. Note that with 100 trials, if p is $\frac{3}{4}$, the probability that Smith wins the bet is nearly 1, while in the case of eight out of ten, it was only about

$\frac{1}{2}$. Thus, in the case of 100 trials, it would be easy to convince both Smith and Jones that whichever one is correct is very likely to win the bet.

Thus, we see that the probability of both types of errors can be made small at the expense of having a large number of experiments.

In applications, it is important to have some estimate of the number of experiments that are necessary to reduce the probabilities of errors to acceptable levels. Assume, for example, that we are trying to decide for an independent trials process whether the true probability is p_0 or p_1. Assume that $p_0 < p_1$. We want to design a test so that the probability of error under either hypothesis is at most *a*. We choose a number *s* so that the area under the normal curve beyond *s* is *a*. We perform *n* experiments. If p_0 is correct, the probability of the number of successes *x* exceeding the expected number np_0 by *s* standard deviations is *a*. That is, if p_0 is correct, then

$$\Pr\left[x > np_0 + s\sqrt{np_0q_0}\right] = a$$

On the other hand, if p_1 is correct, the probability that the number of successes will be more than *s* standard deviations below the expected value of np_1 is also *a*. That is, if p_1 is correct, then

$$\Pr\left[x > np_1 - s\sqrt{np_1q_1}\right] = a$$

Assume, then, that we can choose *n* so that

$$np_0 + s\sqrt{np_0q_0} < np_1 - s\sqrt{np_1q_1}$$

Then we can choose a value *t* greater than the first number but such that $t - 1$ is less than the second number. We accept p_0 if $x \le t - 1$ and p_1 if $x \ge t$. The test will have a probability of error of at most *a* under either hypothesis. We can achieve this inequality if

$$\sqrt{np_0} + s\sqrt{p_0q_0} < \sqrt{np_1} - s\sqrt{p_1q_1}$$

or

$$s\left[\frac{\sqrt{p_0q_0} + \sqrt{p_1q_1}}{p_1 - p_0}\right] < \sqrt{n}$$

or

$$n > s^2 \left[\frac{\sqrt{p_0 q_0} + \sqrt{p_1 q_1}}{p_1 - p_0} \right]^2.$$

For example, in our beer and ale example, assume that $p_0 = .5$ and $p_1 = .75$. We would like to be 90 per cent certain of being correct. Then, we see that $s = 1.3$. Thus, we must have

$$n > (1.3)^2 \left[\frac{\sqrt{.5 \times .5} + \sqrt{.75 \times .25}}{.75 - .5} \right]^2 = 23.5.$$

We would need only a moderate number of experiments, namely, 24. Then $np_0 + s\sqrt{np_0 q_0} = 15.18$ and $np_1 - s\sqrt{np_1 q_1} = 15.24$. Jones is 90 per cent sure that Smith will have fewer than 16 correct guesses, while Smith is 90 per cent sure that he will have more than 15 correct guesses. Thus, we award the bet to Smith if he guesses correctly at least 16 times out of 24 experiments.

Consider, however, the Salk vaccine experiment. In this experiment, we want to test $p_0 = .00025$ against $p_1 = .00050$—that is, whether the vaccine will reduce the incidence of polio from 50 to 25 per 100,000. We would want a great deal of reliability for such a test. Let us choose s so that the probability of error is less than .001. We can have this by choosing $s = 3.1$. Then we must have

$$n \geq (3.1)^2 \times \left[\frac{\sqrt{.00025 \times .99975} + \sqrt{.0005 \times .9995}}{.00025} \right] = 223{,}956.$$

In one of the major parts of the Salk vaccine experiment the vaccine was given to 200,000 students. Of these vaccinated students, 57 contracted polio. In Exercise 10, you are asked to design an experiment to test the hypothesis $p_1 = .00050$ against the hypothesis $p_0 = .00025$.

Exercises

1. Assume that in the beer and ale experiment, Jones agrees to pay Smith if Smith gets at least nine out of ten correct.

(a) What is the probability of Jones paying Smith even though Smith cannot distinguish beer and ale, and guesses?

[*Ans.* .011]

(b) Suppose that Smith can distinguish with probability .9. What is the probability of his not collecting from Jones?

[*Ans.* .264]

2. Suppose that in the beer and ale experiment, Jones wishes the probability to be less than .1 that Smith will be paid if, in fact, he guesses. How many often trials must he insist that Smith gets correct to achieve this?

3. In the analysis of the beer and ale experiment, we assume that the various trials were independent. Discuss several ways that error can enter, because of the non-independence of the trials, and how this error can be eliminated. (For example, the glasses in which the beer and ale were served might be distinguishable).

4. Consider the following two procedures for testing Smith's ability to distinguish beer from ale.

   (a) Four glasses are given at each trial, three containing beer and one ale, and he is asked to pick out the one containing ale. This procedure is repeated ten times. He must guess correctly seven or more times.

   (b) Ten glasses are given to him, and he is told that five contain beer and five ale, and he is asked to name the five which he believes contain ale. He must choose all five correctly.

   In each case, find the probability that Smith establishes his claim by guessing. Is there any reason to prefer one test over the other?

   [*Ans.* (a) .003; (b) .004]

5. A testing service claims to have a method for predicting the order in which a group of freshmen will finish in their scholastic record at the end of college. The college agrees to try the method on a group of five students, and says that it will adopt the method if, for these five students, the prediction is either exactly correct or can be changed into the correct order by interchanging one pair of *adjacent* men in the predicted order. If the method is equivalent

to simply guessing, what is the probability that it will be accepted?

[*Ans.* $\frac{1}{24}$]

6. The standard treatment for a certain disease leads to a cure in $\frac{1}{4}$ of the cases. It is claimed that a new treatment will result in a cure in $\frac{3}{4}$ of the cases. The new treatment is to be tested on ten people having the disease. If seven or more are cured, the new treatment will be adopted. If three or fewer people are cured, the treatment will not be considered further. If the number cured is four, five, or six, the results will be called inconclusive, and a further study will be made. Find the probabilities for each of these three alternatives first, under the assumption that the new treatment has the same effectiveness as the old, and second, under the assumption that the claim made for the treatment is correct.

7. Three upperclassmen debate the intelligence of the freshmen class. One claims that most freshmen (say 90 per cent of them) are intelligent. A second claims that very few (say 10 per cent) of them are intelligent, while a third one claims that a freshman is just as likely to be intelligent as not. They administer an intelligence test to ten freshmen, classifying them as intelligent or not. They agree that the first man wins the bet if eight or more are intelligent, the second if two or fewer, the third in all other cases. For each man, calculate the probability that he wins the bet, if he is right.

[*Ans.* .930, .930, .890]

8. Ten men take a test with ten problems. Each man on each question has probability $\frac{1}{2}$ of being right, if he does not cheat. The instructor determines the number of students who get each problem correct. If he finds on four or more problems there are fewer than three or more than seven correct, he considers this convincing evidence of communication between the students. Give a justification for the procedure. [*Hint:* The table in Figure 4.2 must be used twice, once for the probability of fewer than three or more than seven correct answers on a given problem, and the second time to find the probability of this happening on four or more problems.]

9. An instructor claims that a certain student knows only 70 per cent of the material. The student claims that he knows 85 per cent. Design a test that will settle the argument with probability .9.

[*Ans.* 50 questions, student must get 40 correct answers]

10. Assume that the Salk vaccine is to be given to 225,000 students. It is claimed that the probability of getting polio is $\leq .00025$ if vaccinated and .00050 if not vaccinated. Design a test to decide between these two alternatives. In the actual experiment, there were 28 cases per 100,000 of polio among the 200,000 vaccinated. This would suggest 63 cases in 225,000 students. Would your test establish the claim that the Salk vaccine was effective, if these few cases of polio occurred in the experiment?

Confidence Intervals

Consider n independent trials with probability p for success on each trial. We assume that we do not know p but want to make, on the basis of our observations, some estimate of p. Let a be any number between 0 and 1. Then from Figure 4.6 we can find a number s such that the area under the normal curve beyond s is $a/2$. For example, if $a = .05$ then we can choose $s = 2$. By the central limit theorem, if x is the number of successes, then

$$\Pr\left[\left|\frac{x-np}{\sqrt{npq}}\right| \leq s\right] \approx 1-a.$$

This is the same as saying that

$$\Pr\left[\left|\frac{x/n-p}{\sqrt{pq/n}}\right| \leq s\right] \approx 1-a.$$

Putting $\bar{p} = x/n$, we have

$$\Pr\left[|\bar{p}-p| \leq s\sqrt{pq/n}\right] \approx 1-a.$$

Using the fact that $pq \leq \frac{1}{4}$ for all p, we have

$$\Pr\left[|\bar{p}-p| \leq s/2\sqrt{n}\right] \geq 1-a.$$

Thus, no matter what p is, with probability at least $1 - a$, the true value will not deviate from $\bar{p}$ by more than $s/2\sqrt{n}$. We say then that

$$\bar{p}-\frac{s}{2\sqrt{n}} \leq p \leq \bar{p}+\frac{s}{2\sqrt{n}}$$

with confidence $I - a$. We call the interval

$$\left[\bar{p} - \frac{s}{2\sqrt{n}}, \bar{p} + \frac{s}{2\sqrt{n}}\right]$$

a $100(1 - a)$ per cent confidence interval. For example, the 95 per cent confidence interval requires $s = 2$, and hence is

$$\left[p - \frac{1}{\sqrt{n}}, p + \frac{1}{\sqrt{n}}\right].$$

For example, if in 400 trials a drug is found effective 124 times or .31 of the time, the 95 per cent confidence interval for p is

$$\left[.31 - \frac{1}{20}, .31 + \frac{1}{20}\right],$$

or [.26, .36]. The 99 per cent confidence interval would be found by using $s = 2.6$. This gives

$$\left[.31 - \frac{2.6}{40}, .31 + \frac{2.6}{40}\right],$$

or [.245, .375]. Of course, as we demand more confidence our prediction is more conservative, i.e., the interval is larger.

.63	.83
.59	.79
.67	.87
.65	.85
.59	.79
.65	.85
.61	.81
.51	.71
.61	.81
.57	.77
.58	.78

.59	.79
.61	.81
.60	.80
.68	.88
.68	.88
.66	.86
.56	.76
.60	.80
.63	.83

Figure: 4.8.

It is important to realise that the interval obtained depends upon the value of $\overline{p}$, which in turn depends upon the value of x. Thus, $\overline{p}$ is a chance quantity. We are assuming that the true value p, though unknown, is not a chance quantity. Thus, our confidence interval itself is a chance quantity which may or may not cover the true value p. When we choose a 95 per cent confidence interval we mean that the probability is .95 that the interval will cover the true value p. Thus, by the law of large numbers we expect this to be the case about 95 per cent of the time.

In Figure 4.8 we give the results of computing the 95 per cent confidence intervals based upon several experiments with $n = 100$ trials for a true value of $p = .7$. We carried out this experiment 20 times. It will be noted that in each case the interval does include the true value, though sometimes just barely. We should not have been surprised if in one or two cases it did not.

The use of the inequality $pq \leq \frac{1}{4}$ was for convenience and simplicity of our computations. It results in slightly larger confidence intervals than are necessary for a given confidence level. Without making this approximation, it is possible to transform the first inequality into an inequality about $\overline{p}$ to obtain a more exact confidence interval.

As a second example of confidence intervals consider the following problem. In a small town, lottery tickets numbered from 1 to N are being sold weekly and a prize is given to the person who holds the ticket having the lucky number drawn at random from the numbers from 1 to N. The value of N is not publicly announced, but is the same every week. A man buys lottery tickets for ten weeks, receiving numbers 27, 46, 77, 85, 34, 24, 34, 46, 34, and 89. Before buying a ticket the following week, he wants to estimate his chance of winning, i.e., he wants to estimate N. Of course, he knows that N is at least 89, the highest number that he has drawn.

n	$\frac{1}{(.05)^{1/n}}$
5	1.821
6	1.648
7	1.534
8	1.454
9	1.395
10	1.349

Figure: 4.9

Let us see how we would obtain confidence intervals for the unknown "parameter" N. The man has in effect drawn a number from the N possible numbers n times. Let M be the maximum of the numbers drawn. Then for fixed n and N, M may be considered a chance quantity. For any A,

$$\Pr[M \le A] = \left(\frac{A}{N}\right)^n .$$

As before, let a be any number between 0 and 1. We can choose A so that $A = a^{1/n}N$. Then

M	$M/(.05^{.1})$
100	134
96	129
89	120
88	118
99	133
93	125

97	130
85	114
74	99
99	133
82	110
98	132
97	130
91	122
93	125
96	129
97	130
90	121
91	122
98	132

Figure: 4.10

$$\Pr\left[M \le a^{1/n}N\right] = \frac{\left(a^{1/n}N\right)^n}{N^n} = a$$

or

$$\Pr\left[\frac{M}{a^{1/n}} \le N\right] = a.$$

That is,

$$\Pr\left[N < \frac{M}{a^{1/n}}\right] = 1 - a$$

Since $M \le N$, we can write this as

$$\Pr\left[M \le N < \frac{M}{a^{1/n}}\right] = 1 - a$$

Thus, the interval $\left[M, M / a^{1-n}\right]$ has probability $1 - a$ of covering N and hence is a $100(1 - a)$ per cent confidence interval for N. In any given example, for a 95 per cent confidence interval, we choose $a = .05$ and hence $M/a^{1/n} = 89/(.5)^{1/10} = 120.1$. Hence, the man can be 95 per cent sure that there are at most 120 lottery tickets.

For such calculations, the table in Figure 4.9 is useful.

In Figure 4.10 we have indicated the result of twenty experiments with N equal in each case to 100 and $n = 10$. We have computed the 95 per cent confidence intervals. In this case, we see that one interval does not include the true value of N. Thus, the intervals include the true value of N precisely 95 per cent of the time.

Example: The Fish and Game Department is interested in estimating the number of trout in a pond (which contains only trout). They take out a sample of 1000 fish and mark them. Later, they take another sample of 1600 and find that 120 of them are marked. What is a reasonable estimate for the total number of trout?

Let n be the unknown total. Since 1000 of them were marked, there is probability $p = 1000/n$ that a fish in the second sample will be marked.

The observed fraction is $\overline{p} = 120/1600$, and the 95 per cent confidence interval yields

$$\frac{120}{1600} - \frac{1}{\sqrt{1600}} \le p \le \frac{120}{1600} + \frac{1}{\sqrt{1600}},$$

or

$$\frac{3}{40} - \frac{1}{40} \le p \le \frac{3}{40} + \frac{1}{40},$$

or

$$\frac{1}{20} \le p \le \frac{1}{10}.$$

Hence,

$$\frac{1}{20} \le \frac{1000}{n} \le \frac{1}{10},$$

and we obtain the estimate $10{,}000 \le n \le 20{,}000$.

Exercises

1. A prospective college student visits a college and sits in on a class of 50 students. She notes that there are 39 men and 10 women in the class. She decides to compute the 95 per cent confidence interval for the proportion of women in the school. She will reject the school if this interval excludes the possibility that $\frac{1}{3}$ of the students are women. Does she reject the school for this reason?

2. A young ballplayer in his first season is at bat 400 times and gets 100 hits for a batting average of .250. Find 90 per cent confidence limits for his batting average based upon his first season. Is it reasonable to believe that he may in fact be a .300 batter?

   [*Ans.* [.209, .291]; maybe, next year]

3. A large company has as many as a million accounts. It wishes to estimate the number that are at least three months delinquent in their payments. A thousand accounts are randomly selected and of these it is observed that 30 are at least three months delinquent. Find the 95 per cent confidence limits for the proportion of customers that are at least three months behind in their payments.

4. Opinion pollsters in election years usually poll about 3,000 voters. Suppose that in an election year 51 per cent favour candidate A and 49 per cent favour candidate B in a poll. Construct 95 per cent confidence limits on the true percentage of the population in favour of A.

   [*Ans.* .492, .528]

5. An experimenter has an independent trials process and she has a hypothesis that the true value of p is p_0. She decides to carry out a number of trials, and from the observed $\bar{p}$ calculate the 95 per cent confidence interval of p. She will reject p_0 if it does not fall within these limits. What is the probability that she will reject p_0 when in fact it is correct? Should she accept p_0 if it does fall within the confidence interval?

6. A coin is tossed 100 times and turns up heads 61 times. Using the method of Exercise 5, test the hypothesis that the coin is a fair coin.

   [*Ans.* Reject]

7. In an experiment with independent trials we are going to estimate p by the fraction $\overline{p}$ of successes. We wish our estimate to be within .02 of the correct value with probability .95. Show that 2,500 observations will always suffice. Show that if it is known that p is approximately .1, then 900 observations would be sufficient.

8. In the Weldon dice experiment, 12 dice were thrown 26,306 times and the appearance of a 5 or a 6 was considered to be a success. The mean number of successes observed was, to four decimal places, 4.0524. Is this result significantly different from the expected average number of 4? [*Ans.* Yes]

9. Prove that $pq \leq \frac{1}{4}$. [*Hint:* write $p = \frac{1}{2} + x$]

10. Suppose that out of 1000 persons interviewed 650 said that they would vote for Mr. Big for mayor. Construct the 99 per cent confidence interval for p, the proportion in the city that would vote for Mr. Big.

11. In a pond, 400 fish are marked. If in a subsequent sample of 225 there are 45 marked fish, find the 90 per cent confidence interval for the total number of fish.

12. In a large city each taxi is assigned a number. A man observes the numbers 125, 135, 356, 344, 25, 299, and 320 on seven occasions that he takes a cab. On the basis of this, compute the 95 per cent confidence limits for the number of cabs in the city. If he knows that the number of cabs is a multiple of 100, can he determine the total?

13. Suppose that the man in Exercise 12 takes three more cabs numbered 76, 421, and 211. Can he be 95 per cent sure of the total?

[*Ans.* Yes]

14. In this section, we have approximated confidence limits on p such that $\Pr\left[|\overline{p} - p| \leq s\sqrt{\frac{pq}{n}}\right] = 1 - a$. The expression inside the brackets is equivalent to $(\overline{p} - p)^2 \leq s^2\left(\frac{p(1-p)}{n}\right)$. Substituting equality for inequality, we obtain a quadratic equation which can be solved for p in terms of $\overline{p}$, s, and n. There will be two roots r_1 and r_2, where $r_1 \leq r_2$ and the above inequality will be satisfied

for all p such that $r_1 \leq p \leq r_2$. Use this information to obtain more exact confidence intervals than that obtained by setting $pq = \frac{1}{4}$.

15. A hundred names are picked at random out of a large telephone book. It is found that 70 of these names have eight letters or less. Place 95 per cent confidence limits on the fraction of names in that telephone book containing eight letters or less:

    (a) Using the estimate developed in the text.

    (b) Using the limits developed in exercise 14.

    [*Ans.* .602, .783]

    (c) Suppose we were using the method of Exercise 5 to test the hypothesis that 79 per cent of the names have eight or less letters. Which of the above intervals would be better?

Some Pitfalls

Statistics properly used is a very powerful tool. If it is not properly used, it can lead to incorrect predictions and thereby cause considerable distrust in its methods. We have already mentioned the example of the poll of the *Literary Digest* in the 1936 presidential election between Roosevelt and Landon. In this poll about 10 million postcards were sent to persons whose names were obtained from telephone directories and car registrations. Several million cards were returned, with 40.9 per cent in favour of Franklin Roosevelt. A few weeks later in the actual election, Roosevelt obtained 60.7 per cent of the vote.

.586	.618
.591	.623
.594	.626
.59	.622
.594	.626
.598	.63
.591	.623
.591	.623
.592	.625
.595	.627

.59	.622
.583	.616
.588	.62
.591	.623
.603	.635
.59	.622
.592	.624
.597	.629
.595	.627
.588	.62

Figure: 4.11

There are two obvious flaws in the above procedure. The first, and the one which is normally blamed for the error, is that people who had telephones or cars at that time were not truly representative of the voting population as a whole. The second is the possibility that people change their minds between the time a poll is taken and the election takes place. They may even deliberately tell the poll taker one thing and vote another. Of course, with many millions of people, it is difficult to choose a truly random sample. However, let us assume that there were in fact 60.7 per cent of the people in favour of Roosevelt at the time of the poll and that we could choose a random sample of only 10,000 voters. In Figure 4.11, we indicate the result of simulating 30 such samples and determining the 99 per cent confidence intervals. We see that in every case, we would have picked Roosevelt to win. This is on the basis of only 10,000 samples rather than the millions which led to a wrong answer. Thus, if statistics can be properly used, it is a very powerful tool.

Because of the difficulties indicated above there is still, with some justice, skepticism of polls. However, there is also some danger in refusing to use statistical methods. For example, assume that an all-male college wishes to know the opinion of its alumni on the question of becoming a coeducational institution. Assume that there are 30,000 alumni and in fact 60 per cent are in favour of the college admitting women. This is a situation in which any person not asked could conceivably challenge the poll. Assume then that it is decided to poll by mail *all* the alumni. Also assume that a proportion p of those who favour coeducation will respond to the query and a proportion 2/7 of those who oppose will respond because they feel more strongly about the matter. Then the expected number of yes answers would be 18,000/J and the no answers would be 24,000/7.

Thus, neglecting sampling errors, the vote would be 18,000/7/42,000/7 = .43 in favour, and coeducation would be defeated. On the other hand, as we have seen, a relatively small random sample in which the response of each person sampled was recorded would give a much more reliable indication of the true feelings of the alumni. For example, a sample of 1000 was taken in such a poll and a 95 per cent confidence interval of (.559, .621) was obtained.

As we have indicated previously, a number of precautions had to be taken in the experiment to test the effectiveness of Salk vaccine. First, although initially the incidence of polio was only about 50 per 100,000, there was considerable variability from year to year and from region to region. So a reduction from 50 to 25 in a sample of 100,000 could easily be caused by reasons having nothing to do with the effectiveness of the vaccine. Thus, it was decided to have control groups. In one part of the experiment, a population of students was divided into two groups of about 200,000 each. All were inoculated at the same time.

The first group received the Salk vaccine and the second a harmless and useless salt solution (a "placebo"). The decision as to which students received the real vaccine was made randomly, and the knowledge of whether a student was given the vaccine or the placebo was not made known to the student or to the physician observing the student. The reason for this is that in such experiments knowledge of whether the subject has been treated or not has been found to introduce a bias in the diagnosis and in the behaviour of the subject. As indicated earlier, the test did show a significantly lower rate among those vaccinated. The test led to further development of vaccines and the virtual elimination of polio in the United States.

There have been a large number of statistical studies to determine if smoking is injurious to one's health. It is now widely believed that this is the case. However, the problem of establishing this has been exceedingly difficult, and there are still statisticians who feel that more testing must be done. In the case of the polio vaccine, it was possible to select two groups and randomly give one half the vaccine and the other half a placebo. Random selection eliminates the effect of biases which can creep in, such as differences in age, place of residence, economic status, etc. To do the corresponding experiment for smoking would require one randomly selected group to become heavy smokers and the rest to abstain.

This is clearly not possible, and many of the studies have had to rely on choosing groups in a less random way and studying their smoking and health patterns.

In the exercises some of these methods are briefly mentioned and you are asked to consider possible pitfalls. While statisticians who criticise these tests or refuse to accept their conclusions are often accused of being overly cautious, their criticisms have led to the development of more careful methods of statistical tests in these very difficult areas.

It should be emphasised that demonstrating that more heavy smokers than non-smokers get lung cancer does not demonstrate that smoking is a cause of lung cancer. It seems likely that there will still be controversy about this question until more knowledge is obtained as to what is the essential cause of cancer.

In the *Literary Digest* poll, the particular people that responded to the postcard enquiry was a chance quantity. In effect, the *size* of the sample was random. While pollsters do not intentionally take advantage of this, the results under such circumstances can be distorted. We illustrate this in an extreme case where the experimenter deliberately tries to take advantage of the randomness of the size of the experiment.

Assume that Mr. Esp claims that he has extrasensory perception. An experiment is arranged in which he is to tell, when a card is placed face down, whether it has a circle or a square on it. Of course, we would want to run a large number of experiments, but for the point we are trying to make we can take a small number, say four. If Mr. Esp is just guessing, we can find his expected score (percentage correct) in the usual manner. The tree and tree measure are shown in Figure 4.12, and his expected score is

$$1\times\frac{1}{16}+\frac{3}{4}\times\frac{4}{16}+\frac{1}{2}\times\frac{6}{16}+\frac{1}{4}\times\frac{4}{16}+0\times\frac{1}{16}=\frac{1}{2}.$$

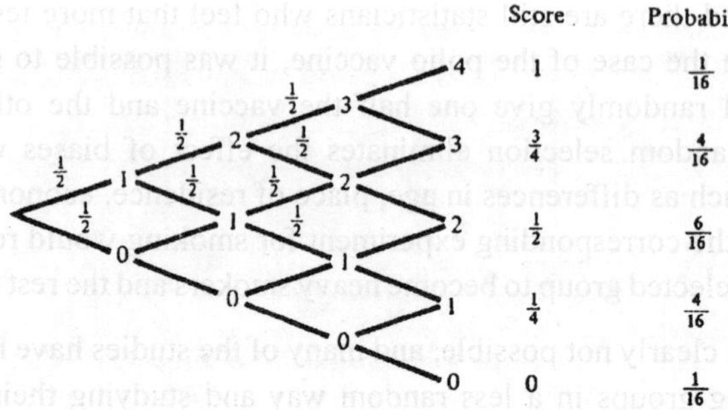

Score	Probability
1	$\frac{1}{16}$
$\frac{3}{4}$	$\frac{4}{16}$
$\frac{1}{2}$	$\frac{6}{16}$
$\frac{1}{4}$	$\frac{4}{16}$
0	$\frac{1}{16}$

Figure: 4.12

Assume now that the experimenter, eager to find a good subject, stops the experiment the first time (if any) that Mr. Esp's score is greater than $\frac{1}{2}$. Then the new tree measure, still assuming guessing, is shown in Figure 4.13. We see now that his expected score is

$$1\times\frac{1}{2}+\frac{2}{3}\times\frac{1}{8}+\frac{1}{2}\times\frac{1}{8}+\frac{1}{4}\times\frac{3}{16}+0\times\frac{1}{16}=\frac{133}{192}=.69,$$

which is considerably better than before.

It is extremely important in designing a statistical test to decide upon the criteria for acceptance or rejection before the test is carried out. Of course, we should not be surprised if we find some unlikely feature of an experiment by looking after the fact for something of small probability.

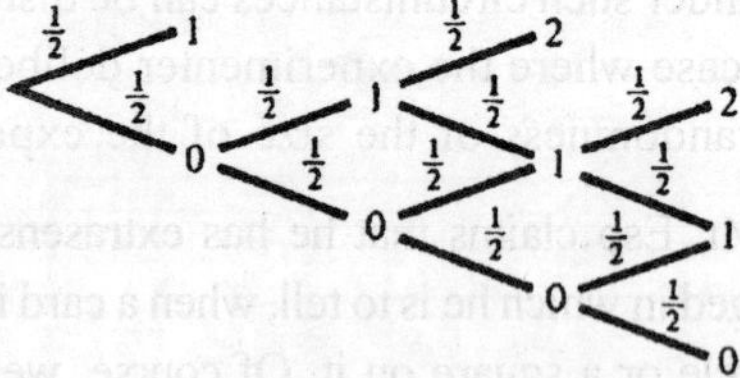

Figure: 4.13

A local expert on probability theory would occasionally be roused from bed at one in the morning to have an excited colleague ask, "What is the probability of being dealt a hand of all hearts in bridge." He would answer, "The same as any other hand," and then go back to sleep.

Exercises

1. In the tests of the Salk vaccine 400,000 students volunteered to be vaccinated. Half were vaccinated and half given placebo. Among these 400,000 people, 199 got polio. Assuming that a person who had polio was equally likely to be in either of the groups, place 99 per cent confidence limits on the number of people in the vaccinated group that got polio. Does the fact that of the 199 reported cases 142 were in the placebo group suggest that the vaccine was effective?
2. Referring to Exercise 1, data was taken also on 340,000 students who did not volunteer to be inoculated. Assuming that these people had the same probability of getting polio as those who received the placebo, place 99 per cent confidence limits on the number of people among this group to get polio. What does the fact that among this group there were 157 polio cases suggest? How might we explain this result?
3. In many of the major studies of smoking and health, the samples are obtained by interviewing whomever happens to be at home when the interviewer calls. This person answers questions relating to everyone in the family over 21. Comment on some possible defects in this method of sampling.
4. In one major study of smoking and health, two groups were compared, one that had lung cancer and another that was chosen by virtue of having similar backgrounds to the group that had cancer. Comment on this technique of sampling.
5. This exercise is designed to show that optional stopping in sampling can significantly change the results. Consider the following game. A box contains five balls, three of which are red and two blue. If a red ball is drawn, we lose a dollar; if a blue ball, we win a dollar.

   (a) Find the expected value of the game if one ball is drawn.

   [*Ans.* – 20 cents]

   (b) Show that the game becomes increasingly unfavourable if two, three, four, or five balls are drawn.

   (c) Show that the game is favourable if you are allowed to stop at any time. Use the following strategy: If the first draw is blue, stop. Otherwise, play until you are even or until all five balls are drawn.

[*Partial Ans:* Value is +20 cents]

6. In a certain college, 25 out of 324 faculty members with Ph.D.'s are women. Nationally approximately 20 per cent of all Ph.D.'s are awarded to women. Test the hypothesis that the faculty members were picked from the national pool without regard to sex.
7. In regarding the significance of Exercise 6 as evidence of discrimination, what other factors would have to be taken into account? For example, is it sufficient to know the present percentage of women among Ph.D.'s?

   And should one know something about the distribution among disciplines?
8. The President of the United States announces a major policy decision. His mail the following week contains 25,000 irate letters and 10,000 favouring his decision. Would it be reasonable to conclude that a majority of people oppose his decision?

Linear Programming

Polyhedral Convex Sets

Recall that an equation containing one or more variables is called an *open statement.* For instance,

$$-2x_1 + 3x_2 = 6 \qquad \text{(a)}$$

is an example of an open statement. If we let $A = (-2, 3)$, $x = \begin{pmatrix} x_1 \\ x_2 \end{pmatrix}$, and $b = 6$, we can write (a) in matrix form as

$$Ax = (-2,3)\begin{pmatrix} x_1 \\ x_2 \end{pmatrix} = (-2, 3) = -2x_1 + 3x_2 = 6 = b.$$

For some two-component vectors x the statement $Ax = b$ is true and for others it is false. For instance, if $x = \begin{pmatrix} 3 \\ 4 \end{pmatrix}$, it is true, since $-2 \cdot 3 + 3 \cdot 4 = 6$; and if $x = \begin{pmatrix} 2 \\ 4 \end{pmatrix}$, it is false, since $-2 - 2 + 3 - 4 = 8$. The set of all two-component vectors x that make the open statement $Ax = b$ true is defined to be the *truth set* of the open statement.

Example 1: In plane geometry it is usual to picture in the plane the truth sets of open statements such as (a). Thus, we can regard each two-component vector x as being the components of a point in the plane in the usual way.

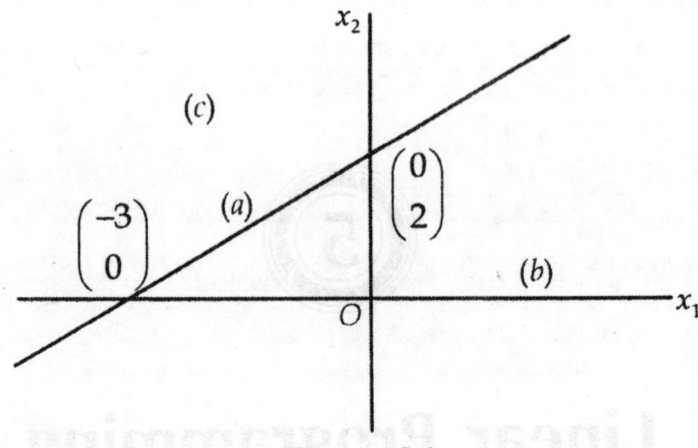

Figure: 5.1

Then the truth set or *locus* (which is the geometric term for truth set) of (a) is the straight line plotted in Figure 5.1. Points on this line may be obtained by assuming values for one of the variables and computing the corresponding values for the other variable. Thus, setting $x_1 = 0$, we find $x_2 = 2$, so that the point $x = \begin{pmatrix}0\\2\end{pmatrix}$ lies on the locus; similarly, setting $x_2 = 0$, we find $x_1 = -3$, so that the point $\begin{pmatrix}-3\\0\end{pmatrix}$ lies on the locus; and so on.

In the same way inequalities of the form $Ax \leq b$ or $Ax < b$ or $Ax \geq b$ or $Ax > b$ are open statements and possess truth sets. And in the case that x is a two-component vector, these can be plotted in the plane.

Example 2: Consider the inequalities (b) $Ax < b$, (c) $Ax > b$, (d) $Ax \leq b$, and (e) $Ax \geq b$, where A, x, and b are as in Example 1. They may be written as

$$-2x_1 + 3x_2 < 6, \tag{b}$$

$$-2x_1 + 3x_2 > 6, \tag{c}$$

$$-2x_1 + 3x_2 \leq 6, \tag{d}$$

$$-2x_1 + 3x_2 \geq 6. \tag{e}$$

Consider (b) first. What points $\begin{pmatrix}x_1\\x_2\end{pmatrix}$ satisfy this inequality? By trial and error we can find many points on the locus. Thus, the point $\begin{pmatrix}1\\2\end{pmatrix}$ is on it, since $-2 \cdot 1 + 3 \cdot 2 = 4 < 6$; on the other hand, the point $\begin{pmatrix} 1 \\ 3 \end{pmatrix}$ is not on the locus, because $-2 \cdot 1 + 3 \cdot 3 = -2 + 9 = 7$, which is not less than 6.

In between these two points we find $\begin{pmatrix} 1 \\ \frac{8}{3} \end{pmatrix}$, which lies on the boundary—that is, on the locus of (a). We note that, starting with $\begin{pmatrix} 1 \\ \frac{8}{3} \end{pmatrix}$ on locus (a), by increasing x_2 we went outside the locus (b); by decreasing x_2 we came into the locus (b) again. This holds in general. Given a point on the locus of (a), by increasing its second coordinate we get more than 6, but by decreasing the second coordinate we get less than 6, and hence the latter gives a point in the truth set of (b). Thus, we find that the locus of (b) consists of all points of the plane below the line (a). The area on one side of a straight line is called an *open half-plane.*

We can apply exactly the same analysis to show that the locus of (c) is the open half-plane above the line (a). This can also be deduced from the fact that the truth sets of statements (a), (b), and (c) are disjoint and have as union the entire plane.

Since (d) is the disjunction of (a) and (b), the truth set of (d) is the union of the truth sets of (a) and (b). Such a set, which consists of an open half-plane together with the points on the line that define the half-plane, is called a *closed half-plane.* Obviously, the truth set of (e) consists of the union of (a) and (c) and therefore is also a closed half-plane.

Frequently we want to assert several different open statements at once—that is, we want to assert the conjunction of several such statements. The easy way to do this is to let A be an $m \times n$ matrix, x an n-component column vector, and b an m-component column vector. Then the statement $Ax \le b$ is the conjunction of the m statements $A_i x \le b_i$, where A_i *is the* ith row of A and b_i is the ith entry of b.

Example 3: A box manufacturer makes small and large boxes from a single kind of cardboard. The small boxes require 2 square feet of cardboard each and the large boxes 3 square feet each. If the manufacturer has 60 square feet of cardboard on hand, what are the possible combinations of small and large boxes that he can make?

In order to set up this problem let x_1 be the number of small boxes and x_2 the number of large boxes to be made. Since it is impossible to make negative numbers of boxes, we have the obvious constraints

$$x_1 \geq 0, \qquad \text{(f)}$$

$$x_2 \geq 0. \qquad \text{(g)}$$

Also, because of the constraint on the total amount of cardboard on hand, we have

$$2x_1 + 3x_2 \leq 60. \qquad \text{(h)}$$

If we now want to state these three inequality constraints simultaneously in the form $Ax \leq b$, we must first change (f) and (g) into $\leq$ constraints.

This can be done by multiplying through by —1, so that (f) becomes $-x_1 \leq 0$ and (g) becomes $-x_2 \leq 0$. If we now define

$$A = \begin{pmatrix} -1 & 0 \\ 0 & -1 \\ 2 & 3 \end{pmatrix}, \quad x = \begin{pmatrix} x_1 \\ x_2 \end{pmatrix}, \quad b = \begin{pmatrix} 0 \\ 0 \\ 60 \end{pmatrix},$$

we see that $Ax \leq b$ is a matrix way of asserting the conjunction of (f), (g), and (h). The truth set of $Ax \leq b$ is the intersection of the three individual truth sets. The truth set of (f) is the right half-plane; the truth set of (g) is the upper half-plane; and the truth set of (h) is the half-plane below and on the line $2x_1 + 3x_2 = 60$. The intersection of these is the triangle (including the sides and corners) shaded in Figure 5.2. The area shaded in Figure 5.2 contains all those and only those points that simultaneously satisfy (f), (g), and (h), or, equivalently, $Ax \leq b$.

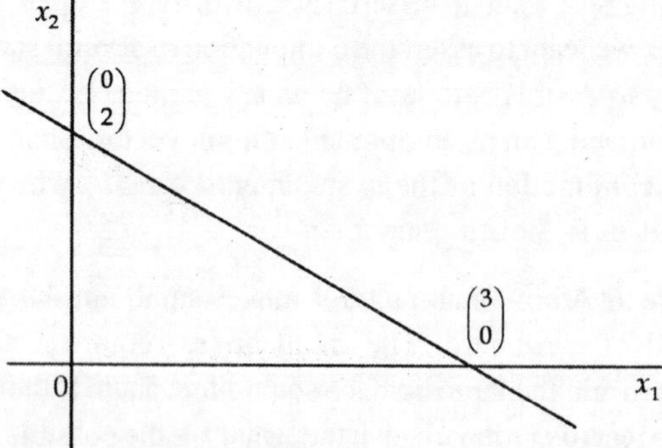

Figure: 5.2

In the examples considered so far we have restricted ourselves to open statements with two variables. Such statements have truth sets that can be sketched in the plane. In the same way, open statements with three variables have truth sets that can be visualised in three-dimensional space. Open statements with four or more variables have truth sets in four or more dimensions, which we can no longer visualise. However, applied problems frequently lead to such statements. Fortunately, we shall develop methods for handling them without having to visualise the truth sets geometrically.

In order to have a notation that will enable us to talk in general about conjunctions of several open statements in any number of dimensions, we shall, for the remainder of this chapter, consider b to be an m-component column vector, x an n-component column vector, and A an $m \times n$ matrix. The ith row of A will be denoted by A_i. Similarly, the ith component of b will be denoted by b_i. Of course, A_i is an n-component row vector and b_i is a number. We shall let x_n denote the set of all n-component column vectors x. Thus, in Example 3 we had $m = 3$ and $n = 2$. A was a 3×2 matrix, x a two-component column vector, and b a three-component column vector. The set of all two-component column vectors x is denoted by X_2.

We now set up some definitions that will be used in the later exposition.

Definition: The truth set of $A_i x = b_i$ is called a hyperplane in X_n. The truth sets of inequalities of the form $A_i x < b_i$ or $A_i x > b_i$ are called open half-spaces, while the truth sets of the inequalities $A_i x \leq b_i$ or $A_i x \geq b_i$ are called closed half-spaces in X_n.

When we assert the conjunction of several open statements, the resulting truth set is the intersection of the truth sets of the individual open statements. Thus, in Example 3 we have the conjunction of $m = 3$ open statements in x_2. In Figure 5.2 we show this geometrically as the intersection of $m = 3$ closed half-spaces (-planes) in $n = 2$ dimensions. Such intersections of closed half-spaces are of special importance.

Definition: The intersection of a finite number of closed half-spaces is a *polyhedral convex set.*

Theorem: Any polyhedral convex set is the truth set of an inequality statement of the form $Ax \leq b$.

Proof: A closed half-space is the truth set of an inequality of the form $A_i x \le b_i$. (An inequality of the form $A_i x \ge b_i$ can be converted into one of this form by multiplying by –1). Now a polyhedral convex set is the truth set of the conjunction of several such statements. Since A is the matrix whose ith row is A_i and b is the column vector with components b_i, then the inequality statement $Ax \le b$ is a succinct way of stating the conjunction of the inequalities $A_1 x \le b_1, \ldots, A_m x \le b_m$. This completes the proof.

The terminology polyhedral *convex* sets is used because these sets are special examples of convex sets. A convex set C is a set such that whenever u and v are points of C, the entire line segment between u and v also belongs to C. This is equivalent to saying that all points of the form $z = au + (1 - a)v$ for $0 \le a \le 1$ belong to C whenever u and v do. In this chapter we shall be concerned primarily with polyhedral convex sets.

Exercises

1. Draw pictures of the truth sets of $Ax \le b$, where A and b are as given below (Construct the truth sets of the individual statements first and then take their intersection).

(a) $A = \begin{pmatrix} 1 & 0 \\ 0 & 1 \\ -2 & -3 \end{pmatrix}, b = \begin{pmatrix} 3 \\ 2 \\ 0 \end{pmatrix}.$

(b) $A = \begin{pmatrix} -2 & -3 \\ -1 & 1 \\ 1 & 1 \end{pmatrix}, b = \begin{pmatrix} -6 \\ 2 \\ 3 \end{pmatrix}.$

(c) $A = \begin{pmatrix} 2 & 3 \\ -1 & 1 \\ 1 & 1 \end{pmatrix}, b = \begin{pmatrix} 6 \\ 2 \\ 3 \end{pmatrix}.$

(d) $A = \begin{pmatrix} 0 & -1 \\ -1 & 0 \\ 1 & 0 \end{pmatrix}, b = \begin{pmatrix} 0 \\ 0 \\ 2 \end{pmatrix}.$

(e) $A = \begin{pmatrix} 1 & 0 \\ -1 & 0 \\ 0 & 1 \\ 0 & -1 \end{pmatrix}, b = \begin{pmatrix} 2 \\ 2 \\ 3 \\ 3 \end{pmatrix}.$

(f) $A = \begin{pmatrix} 3 & 2 \\ 3 & 2 \end{pmatrix}, b = \begin{pmatrix} -6 \\ 6 \end{pmatrix}.$

(g) $A = \begin{pmatrix} -3 & -2 \\ 3 & 2 \end{pmatrix}, b = \begin{pmatrix} -6 \\ 6 \end{pmatrix}.$

(h) $A = \begin{pmatrix} -1 & 1 \\ 1 & 1 \end{pmatrix}, b = \begin{pmatrix} 0 \\ 0 \end{pmatrix}.$

(i) $A = \begin{pmatrix} 1 & 0 \\ -1 & 0 \end{pmatrix}, b = \begin{pmatrix} 2 \\ -5 \end{pmatrix}.$

(j) $A = \begin{pmatrix} -3 & -2 \\ -2 & -3 \\ -1 & 0 \\ 0 & -1 \end{pmatrix}, b = \begin{pmatrix} -6 \\ -6 \\ 0 \\ 0 \end{pmatrix}.$

(k) $A = \begin{pmatrix} -2 & -1 \\ 1 & 0 \\ 0 & 1 \end{pmatrix}, b = \begin{pmatrix} -7 \\ 0 \\ 0 \end{pmatrix}$

2. In the cardboard-box problem of Example 3 consider the following additional constraints:

   (a) "At least as many small as large boxes should be made." Write a constraint involving x_1 and x_2 that expresses this and find A and b. Draw the picture of the resulting convex set.

   [Partial Ans: $-x_1 + x_2 \leq 0$.]

   (b) In addition to the constraints above add a constraint expressing; "at most 20 small boxes should be made." Find A and b and sketch the convex set.

   [Partial Ans: $x_1 \leq 20$.]

3. Of the Polyhedral Convex Sets constructed in Exercise 1, which have a finite area and which have infinite area?

[*Partial Ans:* (c), (d), (f), (h), and (j) are of infinite area; (g) is a line; (i) and (k) are empty.]

4. For each of the following half-planes give an inequality of which it is the truth set.

   (a) The open half-plane above the x_1 axis. [*Ans.* $x_2 > 0$.]

   (b) The closed half-plane on and above the straight line making angles of 45 degrees with the positive x_1 and x_2 axis.

Exercises 5 through 9 refer to a situation in which a retailer is trying to decide how many units of items X and Y he should keep in stock. Let x be the number of units of X and y be the number of units of Y. X costs \$4 per unit and Y costs \$3 per unit.

5. One cannot stock a negative number of units of either X or Y. Write these conditions as inequalities and draw their truth sets.

6. The maximum demand over the period for which the retailer is contemplating holding inventory will not exceed 600 units of X or 600 units of Y. Modify the set found in Exercise 5 to take this into account.

7. The retailer is not willing to tie up more than \$2400 in inventory altogether. Modify the set found in Exercise 6.

8. The retailer decides to invest at least twice as much in inventory of item X as he does in inventory of item Y. Modify the set of Exercise 7.

9. Finally, the retailer decides that he wants to invest \$900 in inventory of item Y. What possibilities are left? [*Ans.* None.]

10. Assume that a pound of meat contains 80 units of protein and 10 units of calcium while a quart of milk contains 15 units of protein and 60 units of calcium. If an adult's minimum daily requirements are 40 units of protein and 30 units of calcium, what consumption quantities of meat and milk will yield at least these minimum daily requirements? A convenient way to summarise the data is by the following data box:

Food	Protein	Calcium
Meat	$80 \dfrac{\text{units protein}}{\text{lb meat}}$	$10 \dfrac{\text{units calcium}}{\text{lb meat}}$

Milk	$15\,\dfrac{\text{units protein}}{\text{qt milk}}$	$60\,\dfrac{\text{units calcium}}{\text{qt milk}}$
Requirements	$40\,\dfrac{\text{units protein}}{\text{day}}$	$30\,\dfrac{\text{units calcium}}{\text{day}}$

(a) Let w_1 be the number of pounds of meat and w_2 be the number of quarts of milk consumed per day, and let $w = (w_1, w_2)$. Write inequality constraints that will solve the above problem. Find A and c so that they can be written $wA \geq c$.

(b) Sketch the set of feasible vectors. Show that it is unbounded (that it has infinite area).

(c) Show that another way of indicating units is as in the data box that follows:

	Protein	Calcium	
Meat	80	10	(per pound)
Milk	15	60	(per quart)
Requirements	40	30	(per day)
	(units)	(units)	

Extreme Points; Maxima and Minima of Linear Functions

In the present section we first discuss the problem of finding the extreme points of a bounded convex polyhedral set. Then we find out how to compute the maximum and minimum values of a linear function defined on such a set.

We use the following notation: the polyhedral convex set C is the truth set of the statement $Ax \leq b$, where A is an $m \times n$ matrix, x is an n-component column vector, and b is an m-component column vector. We let $A_1, A_2, \ldots, A_m$ denote the rows of A. Hence, A_i is an n-component row vector and

$$A = \begin{pmatrix} A_1 \\ A_2 \\ \vdots \\ A_m \end{pmatrix}.$$

The statement $Ax \leq b$ is then the conjunction of the statements

$$A_1x \le b_1,\ A_2x \le b_2, \cdots, A_mx \le b_m.$$

Definition: We shall call the truth set of the statement $A_ix = b_i$ the bounding hyperplane of the half space $A_ix \le b_i$.

Thus, in Figure 5.1 of the preceding section the slanting line (a) is the bounding hyperplane of the half-space (b).

We found in the previous section that a convex set C is the intersection of a finite number of half-spaces. The bounding hyperplanes of these half-spaces that also contain points of C are called *bounding hyperplanes* of C. Thus in example 3 of previous section, the bounding hyperplanes of the polyhedral convex set given there are the three boundary lines of the triangle shaded in Figure 5.2. Note that these lines intersect in pairs in three points, the vertices of the triangle.

Definition: Let C be the polyhedral convex set defined by $Ax \le b,$ where x is an n-component vector. Then a point T is an *extreme* (or *comer) point* of C if it

a. belongs to C, and

b. is the intersection of n bounding hyperplanes of C.

Example 1: Find the extreme points of the polyhedral convex set $Ax \le b,$ where

$$A = \begin{pmatrix} 2 & 3 \\ -2 & -1 \\ 0 & -1 \end{pmatrix},\ x = \begin{pmatrix} x_1 \\ x_2 \end{pmatrix},\ b = \begin{pmatrix} 60 \\ -32 \\ -2 \end{pmatrix}.$$

The corresponding inequalities are:

$$\begin{aligned} 2x_1 + 3x_2 &\le 60, \\ 2x_1 + x_2 &\ge 32, \\ x_2 &\ge 2. \end{aligned}$$

The last two inequalities have been multiplied through by -1, and can be regarded as managerial constraints added to the box-manufacturer problem of Example 3 of preceding section. A sketch of the three half-planes (Figure 5.3) shows that the set of feasible solutions is a triangle. Hence, we can find the extreme points by changing the inequalities to equalities in pairs and solving three sets of simultaneous equations.

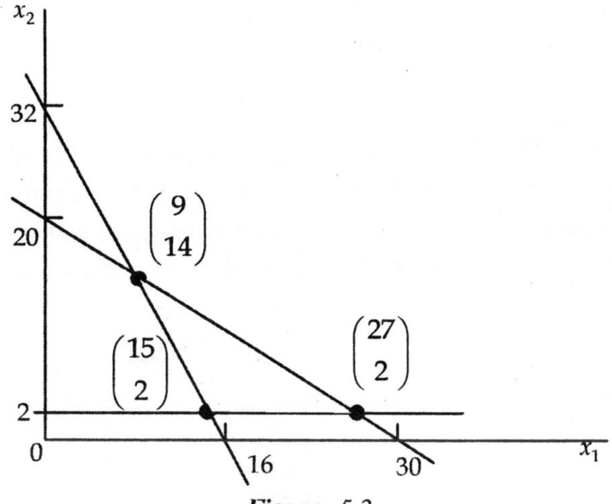

Figure: 5.3

We obtain in this way the points

$$\begin{pmatrix} 9 \\ 14 \end{pmatrix}, \begin{pmatrix} 15 \\ 2 \end{pmatrix}, \begin{pmatrix} 27 \\ 2 \end{pmatrix},$$

which are the Extreme Points of the set.

We can now give an interpretation for the various points of the polyhedral convex set in terms of the system of inequalities. An extreme point lines on ιwo boundaries, which means that two of the inequalities are actually equalities. A point on a side, other than an extreme point, lies on one boundary and hence one inequality is an equality. An interior point of the polygon must, by a process of elimination, correspond to the case where the inequalities are all strict inequalities—that is, not only $\leq$ but $<$ holds.

There is a mechanical (but lengthy) method for finding all the Extreme Points of a polyhedral convex set C defined by $Ax \leq b$. Consider the bounding hyperplanes $A_1x = b_1, \ldots, A_mx = b_m$ of the half-spaces that determine C. Select a subset of n of these hyperplanes and solve their equations simultaneously. If the result is a unique point x^0, then (and only then) check to see whether x^0 belongs to C. If it does, by the above definition, x^0 is an extreme point of C. Moreover, all Extreme Points of C can be found in this manner.

Example 2: Let

$$A = \begin{pmatrix} -1 & 0 \\ 0 & -1 \end{pmatrix} \text{ and } b = \begin{pmatrix} 0 \\ 0 \end{pmatrix}.$$

Then the polyhedral convex set C defined by $Ax \leq b$ is the first quadrant of the x_1, x_2 plane, shaded in Figure 5.4. The only extreme point is the origin, which is the intersection of the lines $x_1 = 0$ and $x_2 = 0$. This is an example of an *unbounded* polyhedral convex set.

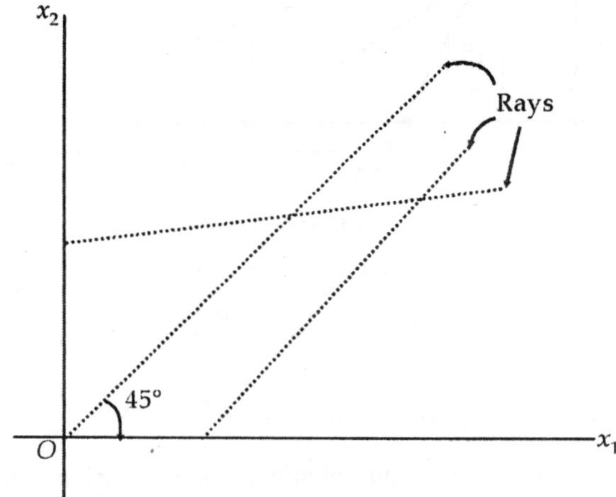

Figure: 5.4

Notice that the set C in Example 2 contains the *ray* or half-line that starts at the origin of coordinates and extends upward to the right making a 45-degree angle with the axes. This ray is dotted in Figure 5.4. Of course, this set also contains many other rays; two others are shown in the figure.

We shall say that a polyhedral convex set is *bounded* if it does not contain a ray. A set, such as the one in Figure 5.4, that does contain rays will be called *unbounded.* For simplicity we shall restrict our discussion in most of this chapter to bounded convex sets.

Example 3: Consider the box-manufacturer problem of Example 3 of preceding section, and suppose that the manufacturer makes a profit of $1 on small and $2 on large boxes. Hence, if he makes x_1 small and x_2 large boxes, his profit function is $x_1 + 2x_2$, and the inequalities limiting the choice of x_1 and x_2 are given in Example 1. What is the most and what the least profit he can make?

We must find the maximum and the minimum value of $x_1 + 2x_2$ for point (x_1, x_2) in the triangle shaded in Figure 5.3. Let us first try the extreme points. At (15, 2) we have a profit of 19, at (27, 2) a profit of 31, and at (9, 14) a profit of 37. The last extreme point is most profitable. But what can we say about the remainder of the triangle? If we start at (9, 14) and try to move to other points in the triangle, the best thing to do is to move along the bounding hyperplane $2x_1 + 3x_2 = 60$, since in this way we can get the most favourable trade-off between x_1 and x_2. However, for each unit we decrease x_2 along this line we can increase x_1 by only $\frac{3}{2}$ units, with a net loss of profit. Hence, the maximum profit is taken on at the extreme point (9, 14). A similar argument shows that the minimum profit is taken on at the extreme point (15, 2). Thus, for this example the maximum and minimum profits are observed at extreme points. We shall show that this is true in general.

Given a convex polyhedral set C and a linear function

$$cx = c_1x_1 + c_2x_2 + \cdots + c_nx_n,$$

where $c = (c_1, c_2, \ldots, c_n)$, we want to show in general that the maximum and minimum values of the function cx always occur at extreme points of C. We shall carry out the proof for the planar case in which $n = 2$, but our results are true in general.

First, we shall show that the values of the linear function $c_1x_1 + c_2x_2$ on any line segment lie between the values the function has at the two endpoints (possibly equal to the value at one endpoint). We represent the points as column vectors $\begin{pmatrix} x_1 \\ x_2 \end{pmatrix}$ and then we see that our linear function is represented by the row vector (c_1, c_2). Let the endpoints of the segment be

$$p = \begin{pmatrix} x_1' \\ x_2' \end{pmatrix} \text{ and } q = \begin{pmatrix} x_1'' \\ x_2'' \end{pmatrix}.$$

We have seen in Figure 2.4 that the points in between p and q can be represented as $tp + (1 - t)q$, with $0 \leq t \leq 1$. If the values of the function at the points p and q are P and Q, respectively (assume that $P \geq Q$), then at a point in between the value will be $tP + (1 - t)Q$, since the function is linear. This value can also be written as

$$tP + (l - t)Q = Q + (P - Q)t,$$

which (for $0 \leq t \leq 1$) is at least Q and at most P.

We are now in a position to prove the result illustrated in Example 3.

Theorem: A linear function cx defined over a convex polyhedral set C takes on its maximum (and minimum) value at an extreme point of C.

Proof: The proof of the theorem is illustrated in Figure 5.5. We shall suppose that at the extreme point p the function takes on a value P greater than or equal to the value at any other extreme point, and at the extreme point q it takes on its smallest extreme-point value, Q.

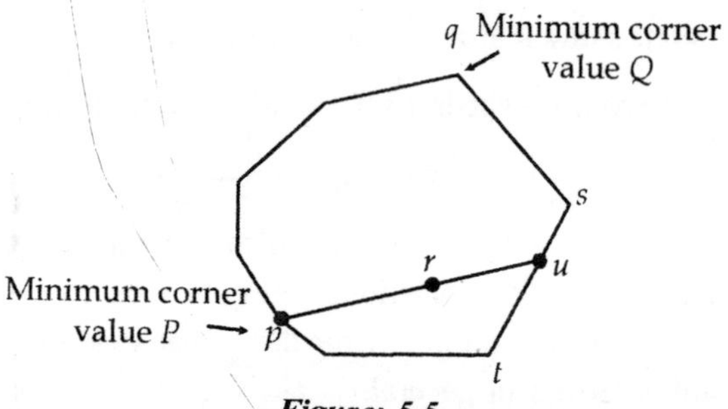

Figure: 5.5

Let r be any point of the polygon. Draw a straight line between p and r and continue it until it cuts the polygon again at a point u lying on an edge of the polygon, say the edge between the corner points s and t (The line may even cut the edge at one of the points s and t; the analysis remains unchanged). By hypothesis the value of the function at any corner point must lie between Q and P. By the above result the value of the function at u must lie between its values at s and t, and hence must also lie between Q and P. Again by the above result the value of the function at r must lie between its values at p and u, and hence must also lie between Q and P. Since r was any point of the polygon, our theorem is proved.

Suppose that in place of the linear function $c_1x_1 + c_2x_2$ we had considered the function $c_1x_1 + c_2x_2 + k$. The addition of the constant k merely changes every value of the function, including the maximum and minimum values of the function, by that amount. Hence, the analysis of where the maximum and minimum values of the function are taken on is unchanged. Therefore, we have the following theorem.

Theorem: The function $cx + k$ defined over a convex polyhedral set C takes on its maximum (and minimum) value at an extreme point of C.

A method of finding the maximum or minimum of the function $cx + k$ defined over a convex set C is then the following: Find the extreme points of the set; there will be a finite number of them; substitute the coordinates of each into the function; the largest of the values so obtained will be the maximum of the function and the smallest value will be the minimum of the function. The method was illustrated previously in Example 3.

Exercises

1. Consider the cardboard-box problem of Exercise 2 of preceding section. Assuming that both constraints stated in (a) and (b) are in effect and the profit function is $x_1 + 2x_2$, find the extreme point (or points) that give maximum and minimum profit.
2. Rework Exercise 1 with profit function $2x_1 + 3x_2$. Show that in this case there is more than one solution for maximum profit.
3. Consider the diet problem of Exercise 10 of preceding section. Suppose that meat costs \$1 per pound and milk costs 30 cents per quart. Find the lowest-cost diet that will meet minimum requirements.

   [*Ans.* $w = \left(\frac{13}{31}, \frac{40}{93}\right)$, cost is $\$\frac{17}{31}$.]
4. The owner of an oil truck with a capacity of 500 gallons hauls gasoline and oil products from city to city. On any given trip he wishes to load his truck with at least 200 gallons of regular gasoline, at least 100 gallons of high-test gasoline, and at most 150 gallons of kerosene. Assuming that he always fills his truck to capacity, find the convex set of ways that he can load his truck. Interpret the extreme points of the set. [*Hint:* There are four extreme points].

5. An advertiser wishes to sponsor a half-hour television comedy and must decide on the composition of the show. The advertiser insists that there be at least three minutes of commercials, while the television network requires that the commercial time be limited to at most 15 minutes. The comedian refuses to work more than 22 minutes each half-hour show. If a band is added to play while neither the comedian nor the commercials are on, construct the convex set C of possible assignments of time to the comedian, the commercials, and the band that use up the 30 minutes. Find the extreme points of C.

   [*Ans.* If x_1 is the comedian time, x_2 the commercial time, and $30 - x_1 - x_2$ the band time, the extreme points are

$$\begin{pmatrix}0\\3\end{pmatrix}, \begin{pmatrix}22\\3\end{pmatrix}, \begin{pmatrix}22\\8\end{pmatrix}, \begin{pmatrix}15\\15\end{pmatrix} \text{ and } \begin{pmatrix}0\\15\end{pmatrix}.]$$

6. In Exercise 4 suppose that the oil truck operator gets 3 cents per gallon for delivering regular gasoline, 2 cents per gallon for high-test, and 1 cent per gallon for kerosene. Write the expression that gives the total amount he will get paid for each possible load that he carries. How should he load his truck in order to earn the maximum amount?

   [*Ans.* He should carry 400 gallons of regular gasoline, 100 gallons of high test, and no kerosene.]

7. In Exercise 6, if he gets 3 cents per gallon for regular and 2 cents per gallon for high-test gasoline, how high must his payment for kerosene become before he will load it on his truck in order to make a maximum profit?

   [*Ans.* He must get paid at least 3 cents per gallon of kerosene.]

8. In Exercise 5 let x_1 be the number of minutes the comedian is on and x_2 the number of minutes the commercial is on the programme. Suppose the comedian costs $200 per minute, the commercials cost $50 per minute, and the band is free. How should the advertiser choose the composition of the show in order that its costs be a minimum?

9. Consider the polyhedral convex set P defined by the inequalities

$$-1 \le x_1 \le 4,$$
$$0 \le x_2 \le 6.$$

Find four different sets of conditions on the constants a and b that the function $F(x) = ax_1 + bx_2$ should have its maximum at one and only one of the four corner points of P. Find conditions that F should have its minimum at each of these points.

[*Ans.* For example, the maximum is at $\begin{pmatrix} 4 \\ 6 \end{pmatrix}$ if $a > 0$ and $b > 0$.]

10. A well-known nursery rhyme goes, "Jack Sprat could eat no fat, his wife could eat no lean...." Suppose Jack wishes to have at least one pound of lean meat per day, while his wife (call her Jill) needs at least .4 pound of fat per day. Assume they buy only beef having 10 per cent fat and 90 per cent lean, and pork having 40 per cent fat and 60 per cent lean. Jack and Jill want to fulfil their minimal diet requirements at the lowest possible cost.

    (a) Let x be the amount of beef and y the amount of pork they purchase per day. Construct the convex set of points in the plane representing purchases that fulfil both persons' minimum diet requirements.

    (b) Suggest necessary restrictions on the purchases that will change this set into a convex polygon.

    (c) If beef costs \$1 per pound, and pork costs 50 cents per pound, show that the diet of least cost has only pork, and find the minimum cost.

    [*Ans.* \$.83.]

    (d) If beef costs 75 cents and pork costs 50 cents per pound, show that there is a whole-line segment of solution points and find the minimum cost.

    [*Ans.* \$.83.]

    (e) If beef and pork each cost \$1 a pound, show that the unique minimal cost diet has both beef and pork. Find the minimum cost.

    [*Ans.* \$1.40.]

    (f) Show that the restriction made in part (b) did not alter the answers given in (c)-(e).

11. In Exercise 10(d) show that for all but one of the minimal-cost diets Jill has more than her minimum requirement of fat, while Jack always gets exactly his minimal requirement of lean. Show that all but one of the minimal-cost diets contains some beef.

12. In Exercise 10(e) show that Jack and Jill each get exactly their minimal requirements.

13. In Exercise 10 if the price of pork is fixed at \$1 a pound, how low must the price of beef fall before Jack and Jill will eat only beef?

[*Ans.* \$.25.]

14. In Exercise 10 suppose that Jack decides to reduce his minimal requirement to .6 pound of lean meat per day. How does the convex set change? How do the solutions in 3(c), (d), and (e) change?

Linear Programming Problems

An important class of practical problems are those that require the determination of the maximum or the minimum of a linear function $cx + k$ defined over a polyhedral convex set of points C. We illustrate these so-called *linear programming problems* by means of the following examples.

Example 1: An automobile manufacturer makes automobiles and trucks in a factory that is divided into two shops. Shop 1, which performs the basic assembly operation, must work 5 man-days on each truck but only 2 man-hours on each automobile. Shop 2, which performs finishing operations, must work 3 man-hours for each automobile or truck that it produces. Because of men and machine limitations shop 1 has 180 man-hours per week available while shop 2 has 135 man-hours per week. If the manufacturer makes a profit of \$300 on each truck and \$200 on each automobile, how many of each should he produce to maximise his profit?

Before proceeding, let us summarise the problem in the *data box* of Figure 5.6 (The term *data box* is due to A. W. Tucker). Notice that the numbers introduced above appear in the data box with their physical dimensions attached. When doing dimensional analysis, in the sense of physics, we may manipulate these dimension quantities just like algebraic quantities. We can obtain interpretations for dual variables by means of dimensional analysis. The reader is strongly advised to set up a similar data box for every linear programming example he works.

An alternate and slightly more elegant way of indicating the units in the data box is shown in Figure 5.7. The reader should compare it with Figure 5.6 to see the correspondence between them. When in doubt, use the more explicit indications of Figure 5.6.

	Trucks	Autos	Capacities
Shop 1	$5\dfrac{S1-\text{manhr}}{\text{truck}}$	$2\dfrac{S1-\text{manhr}}{\text{auto}}$	$180\dfrac{S1-\text{manhr}}{\text{week}}$
Shop 2	$3\dfrac{S2-\text{manhr}}{\text{truck}}$	$3\dfrac{S2-\text{manhr}}{\text{auto}}$	$135\dfrac{S2-\text{manhr}}{\text{week}}$
Profits	$300\dfrac{\$}{\text{truck}}$	$200\dfrac{\$}{\text{auto}}$	

Figure: 5.6

	Trucks	Autos	Capacities	
Shop 1	5	2	180	(Man-hours)
Shop 2	3	3	135	(Man-hours)
Profits	300	200		($)
	(per truck)	(per auto)	(per week)	

Figure: 5.7

A dimensional fraction such as "S1-manhr/truck" is read "shop 1 man-hours per truck." Suppose we now introduce two variables x_1 with dimensions "trucks/week," which will become the number of trucks per week we should produce, and x_2 with dimensions "autos/week." Then the first constraint of the data box of Figure 5.6 becomes:

$$\left(5\frac{\text{S1-manhr}}{\text{truck}}\right)\left(x_1\frac{\text{trucks}}{\text{week}}\right)+\left(2\frac{\text{S1-manhr}}{\text{auto}}\right)\left(x_2\frac{\text{autos}}{\text{week}}\right)<180\frac{\text{S1-mahr}}{\text{week}}.$$

Now, by cancelling the common term "truck" from numerator and denominator of the first term, and similarly cancelling the common dimension "auto" from the numerator and denominator of the second term, we see that the resulting dimensions of each term are "Sl-manhr/ week"—the same as the dimensions of the capacity term on the right-hand side of the inequality. A similar dimensional analysis can be carried out for the second capacity constraint. Dropping dimensions, we have the following restrictions:

$$5x_1+2x_2\leq 180,$$
$$3x_1+3x_2\leq 135,$$

together with the obviously necessary non-negative constraints $x_1\geq 0$ and $x_2\geq 0$.

Subject to these constraints we want to maximise the profit function:

$$\left(300\frac{\$}{\text{truck}}\right)\left(x_1\frac{\text{trucks}}{\text{week}}\right)+\left(200\frac{\$}{\text{auto}}\right)\left(x_2\frac{\text{autos}}{\text{week}}\right).$$

Cancelling out the common terms, we see that the dimensions of this function are simply "\$/week."

In order to state the problem as a linear programming problem we define the quantities:

$$A = \begin{pmatrix} 5 & 2 \\ 3 & 3 \end{pmatrix}, \; b = \begin{pmatrix} 180 \\ 135 \end{pmatrix}, \text{ and } c = (300, 200),$$

which are immediately evident from the data boxes in Figure 5.6 and 5.7. Then our problem is:

Maximum Problem: Determine the vector $x = \begin{pmatrix} x_1 \\ x_2 \end{pmatrix}$ so that the weekly profit, given by the quantity cx, is a maximum subject to the inequality constraints $Ax \le b$ and $x \ge 0$. The inequality constraints insure that the weekly number of available man-hours is not exceeded and that non-negative quantities of automobiles and trucks are produced.

The graph of the convex set of possible x vectors is pictured in Figure 5.8. This is a problem of the kind discussed in the previous section.

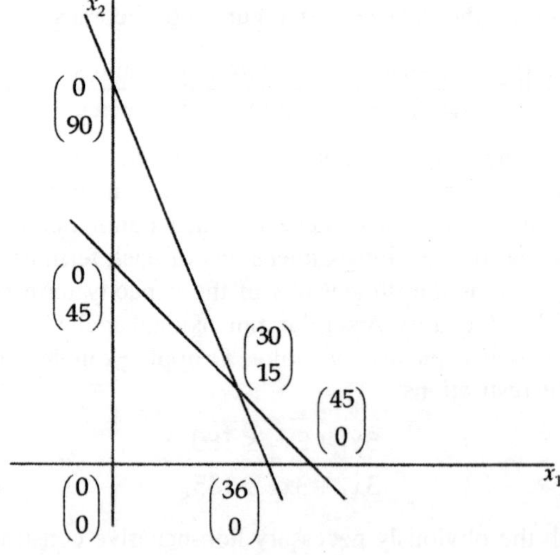

Figure: 5.8

The extreme points of the convex set C are

$$T_1 = \begin{pmatrix} 0 \\ 0 \end{pmatrix}, \; T_2 = \begin{pmatrix} 36 \\ 0 \end{pmatrix}, \; T_3 = \begin{pmatrix} 0 \\ 45 \end{pmatrix}, \text{ and } T_4 = \begin{pmatrix} 30 \\ 15 \end{pmatrix}.$$

Following the solution procedure outlined in the previous section, we test the function $cx = 300x_1 + 200x_2$ at each of these extreme points. The values taken on are 0, 10,800, 9000, and 12,000. Thus, the maximum weekly profit is $12,000, achieved by producing 30 trucks and 15 automobiles per week.

Example 2: A mining company owns two different mines that produce a given kind of ore. The mines are located in different parts of the country and have different production capacities. After crushing, the ore is graded into three classes: high-grade, medium-grade, and low-grade ores. There is some demand for each grade of ore. The mining company has contracted to provide a smelting plant with 12 tons of high-grade, 8 tons of medium-grade, and 24 tons of low-grade ore per week. It costs the company $200 per day to run the first mine and $160 per day to run the second. However, in a day's operation the first mine produces 6 tons of high-grade, 2 tons of medium-grade, and 4 tons of low-grade ore, while the second mine produces daily 2 tons of high-grade, 2 tons of medium-grade, and 12 tons of low-grade ore. How many days a week should each mine be operated in order to fulfil the company's orders most economically?

Before proceeding, we again summarise the problem in the data boxes of Figures 5.9 and 5.10. The reader should compare these two figures to see the correspondence between them.

	High-grade Ore HG	Medium-grade Ore MG	Low-grade Ore LG	Cost
Mine 1	$6\dfrac{\text{tons-HG}}{\text{M1-day}}$	$2\dfrac{\text{tons-MG}}{\text{M1-day}}$	$4\dfrac{\text{tons-LG}}{\text{M1-day}}$	$200\dfrac{\$}{\text{M1-day}}$
Mine 2	$2\dfrac{\text{tons-HG}}{\text{M2-day}}$	$2\dfrac{\text{tons-MG}}{\text{M2-day}}$	$12\dfrac{\text{tons-MG}}{\text{M2-day}}$	$160\dfrac{\$}{\text{M2-day}}$
Requirements	$12\dfrac{\text{tons-HG}}{\text{week}}$	$8\dfrac{\text{tons-MG}}{\text{week}}$	$24\dfrac{\text{tons-LG}}{\text{week}}$	

Figure 5.9

	High-grade Ore	Medium-grade Ore	Low-grade Ore	Cost	
Mine 1	6	2	4	200	(per day)
Mine 2	2	2	12	160	(per day)
Requirements	12	8	24		(per week)
	(tons)	(tons)	(tons)	($)	

Figure: 5.10

Let $v = (v_1, v_2)$ be the two-component row vector whose component v_1 gives the number of days per week that mine 1 operates and v_2 gives the number of days per week that mine 2 operates. If we define the quantities

$$A=\begin{pmatrix} 6 & 2 & 4 \\ 2 & 2 & 12 \end{pmatrix}, c=(12,8,24), \text{ and } b=\begin{pmatrix} 200 \\ 160 \end{pmatrix},$$

which are immediately evident from the data box of Figure 5.9, we can state the problem above as a minimum problem.

Minimum Problem: Determine the vector v so that the weekly operating cost, given by the quantity vb, is a minimum subject to the inequality restraints $uA \geq c$ and $v \geq 0$. The inequality restraints insure that the weekly output requirements are met and the limits on the components of v are not exceeded.

This is a minimum problem of the type discussed in detail in the preceding section. In Figure 5.11 we have graphed the convex polyhedral set C defined by the inequalities $vA \geq c$.

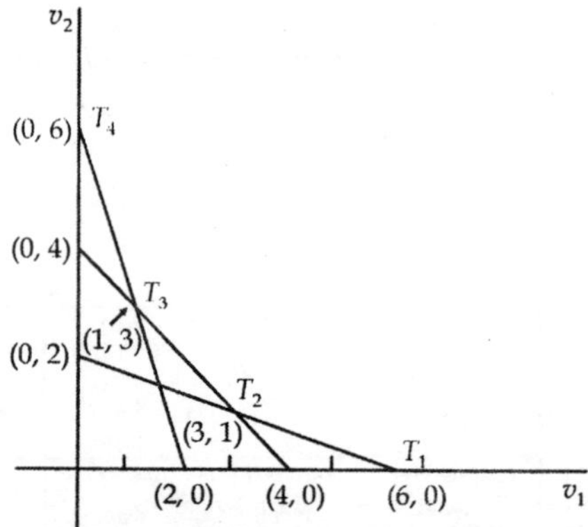

Figure: 5.11

The extreme points of the convex set C are

T_1 = (6, 0), T_2 = (3, 1), T_3 = (1, 3), T_4 = (0, 6).

Testing the function $vb = 200v_1 + 160v_2$ at each of these extreme points, we see that it takes on the values 1200, 760, 680, and 960, respectively. We see that the minimum operating cost is \$680 per week and it is achieved at T_3—that is, by operating the first mine one day a week and the second mine three days a week.

Observe that if the mines are operated as indicated, then the combined weekly production will be 12 tons of high-grade ore, 8 tons of medium-grade ore, and 40 tons of low-grade ore. In other words, for this solution low-grade ore, is overproduced. If the company has no other demand for the low-grade ore, then it must discard 16 tons of it per week in this minimum-cost solution of its production problem.

Example 3: As a variant of Example 2, assume that the cost vector is in other words, the first mine now has a lower daily cost than the second.

$$b = \begin{pmatrix} 160 \\ 200 \end{pmatrix};$$

By the same procedure as above we find that the minimum cost level is again \$680 and is achieved by operating the first mine three days a week and the second mine one day a week. In this solution 20 tons of high-grade ore, instead of the required 12 tons, are produced, while the requirements of medium-and low-grade ores are exactly met. Thus, 8 tons of high-grade ore must be discarded per week.

Example 4: As another variant of Example 2, assume that the cost vector is

$$b = \begin{pmatrix} 200 \\ 200 \end{pmatrix};$$

in other words, both mines have the same production costs. Evaluating the cost function vb at the extreme points of the convex set, we find costs of \$1200 on two of the extreme points (T_1 and T_4) and costs of \$800 on the other two extreme points (T_2 and T_3). Thus, the minimum cost is attained by operating either one of the mines three days a week and the other mine one day a week. But there are other solutions, since if the minimum is taken on at two distinct extreme points it is also taken on at each of the points on the line segment between. Thus, any vector v where $1 \leq v_1 \leq 3$, $1 \leq v_2 \leq 3$, and $v_1 + v_2 = 4$ also gives a minimum-cost solution. For example, each mine could operate two days a week.

It can be shown that for any solution v with $1 < v_1 < 3$, $1 < v_2 < 3$, and $v_1 + v_2 = 4$, both high-grade and low-grade ore are over produced.

Exercises

1. In Example 1, assume that profits are \$200 per truck and \$300 per automobile. What should the factory now produce for maximum profit?
2. In Example 4, show that both high and low-grade ore are overproduced for solution vectors v with $1 < v_1 < 3$, $1 < v_2 < 3$, and $u_1 + v_2 = 4$.
3. A manufacturer has two machines, M_1 and M_2, which he uses to manufacture two products, P_1 and P_2. To produce one unit of P_1, three hours of time on M_1 and six hours on M_2 are needed. And to produce one unit of P_2 takes six hours on M_2 and five hours on M_2. Each machine can run a maximum of 2100 hours per year. If the manufacturer sells product P_1 for a net profit of \$40 and P_2 for a net profit of \$50 each, what production mix shall he produce to maximise his total profit?

   (a) Set up the data box for the problem, marking the dimensions of all numbers.

   (b) Find *A, b,* and *c.*

   (c) Draw the set of possible production vectors and find the optimum profit point.

   [Ans. $x^0 = \begin{pmatrix} 100 \\ 300 \end{pmatrix}$ with yearly profit of \$19,000.]

4. Two breakfast cereals, Krix and Kranch, supply varying amounts of vitamin B and iron; these are listed together with one-third of the daily minimum requirements (MDR) in the table below:

Cereal	*Vitamin B*	*Iron*
Krix	.15 mg/oz	1.67 mg/oz
Kranch	.10 mg/oz	3.33 mg/oz
$\frac{1}{3}$ MDR	.12 mg/day	2.0 mg/day

Krix costs 8 cents an ounce and Kranch 10 cents an ounce. How can we satisfy $\frac{1}{3}$ MDR at minimum cost?

(a) Let v_1 be the amount of Krix eaten and v_2 the amount of Kranch eaten. Write a minimising linear programming problem for the above. Set up the data box and find A, b, and c.

(b) Draw the convex set of possible amounts eaten defined by the inequalities in (a).

(c) What is the lowest-cost feasible diet?

[*Ans.* $v^0 = (.6, .3)$ with cost 7.8 cents.]

5. A farmer owns a 200-acre farm and can plant any combination of two crops I and II. Crop I requires 1 man-day of labour and $10 of capital for each acre planted, while crop II requires 4 man-days of labour and $20 of capital for each acre planted. Crop I produces $40 of net revenue per acre and crop II $60. The farmer has $2200 of capital and 320 man-days of labour available for the year. What is the optimal planting strategy?

[*Ans.* $x^0 = \begin{pmatrix} 180 \\ 20 \end{pmatrix}$ with $8400 revenue.]

6. In Exercise 5 assume that the revenue from crop II is $90 per acre.

(a) Find the new maximum-revenue scheme, and show that now the best thing for the farmer to do is to leave 30 acres unplanted.

(b) Explain why the farmer should leave part of his land fallow in this case.

7. Suppose that a pound of meat contains 1 unit of carbohydrates, 3 units of vitamins, and 12 units of proteins and costs $1. Suppose also that one pound of cabbage contains 3, 4, and 1 units of these items, respectively, and costs 25 cents per pound. If these are the only foods available and the minimum daily requirements are 8 units of carbohydrates, 19 units of vitamins, and 7 units of protein, what is the minimum-cost diet?

[*Ans.* $v^0 = (.2, 4.6)$ with cost $1.35.]

8. Suppose that the minimum-cost diet found in Exercise 7 is unpalatable. In order to increase its palatability, add a constraint requiring that at least a half pound of meat be eaten, and resolve the problem. How much is the cost of the minimum-cost diet increased owing to this palatability constraint? [*Ans.* $.24.]

9. In Exercise 8 suppose that we add a different kind of palatability constraint—namely, that at most two pounds of cabbage be eaten.

Now how much is the cost of the minimum-cost diet increased?

[*Ans.* \$2.82.]

10. A manufacturer produces two types of bearings, *A* and *B*, utilising three types of machines; lathes, grinders, and drill presses. The machinery requirements for one unit of each product, in hours, are expressed in the following table:

Bearing	Machine		
	Lathe	Grinder	Drill Press
A	.01	.03	.03
B	.02	.01	.015
Weekly machine capacity(hr)	400	450	480

He makes a Profit of 10 cents per type A bearing and 15 cents per type B bearing. What should his weekly production of each bearing be in order to maximise his profits?

[*Ans.* $x = \begin{pmatrix} 8000 \\ 16{,}000 \end{pmatrix}$ with weekly profits of \$3200.]

Dual Problem

As the examples of the preceding sections have shown, some linear programming problems are maximising and some are minimising. Thus, we might be interested in maximising profits, production, or market share—or we might want to minimise costs, completion times, or raw-material usage. We shall show that to each maximising problem there is a well-defined minimising problem that uses the same data and whose solution has important mathematical implications concerning the original maximising problem. Similarly, to each minimising problem there is a well-defined maximising problem that uses the same data and is similarly related.

First, we recall that every linear programming problem can be put into one of the two following forms:

$$\left.\begin{aligned} &\text{Minimise } cx \\ &\text{subject to } Ax \le b \\ &\qquad\qquad x \ge 0 \end{aligned}\right\} \text{the MINIMUM problem.} \tag{1}$$

or

$$\left.\begin{array}{ll}\text{Minimise} & vb \\ \text{subject to} & vA \geq c \\ & v \geq 0\end{array}\right\}\text{the MINIMUM problem.} \qquad (2)$$

If the components of A, b, c are the same, then the two problems (1) and (2) are called *dual linear programming problems.* Every linear programming problem, whether of the maximum or minimum type, has a dual that can be formally stated as above. The dual of a given problem frequently has important economic meaning and always has mathematical significance.

To set up a *maximum problem* proceed as follows: Let the variables to be determined be $x_1,\ldots, x_n$; set up the data box as in Figure 5.12, with the x-variables appearing as labels on the top of the box. It then follows that, taking A, b, and c from the data box, the maximum problem is in form (1) above.

x_1	x_2	$\cdots$	x_n	
a_{11}	a_{12}	$\cdots$	a_{1n}	b_1
a_{21}	a_{22}	$\cdots$	a_{2n}	b_2
$\cdot$	$\cdot$	$\cdots$	$\cdot$	$\cdot$
a_{m1}	a_{m2}	$\cdots$	a_{mn}	b_m
c_1	c_2	$\cdots$	c_n	

Figure: 5.12

To set up a *minimum problem* proceed as follows: Let the variables to be determined by $v_1,\ldots,v_m$; set up the data box as in Figure 5.13 with the u-variables appearing as labels to the left of the box. It then follows that, taking A, b, and c from the data box, the minimum problem is in form (2) above.

v_1	a_{11}	a_{12}	$\cdots$	a_{1n}	b_1
v_2	a_{21}	a_{22}	$\cdots$	a_{2n}	b_2
$\cdot$	$\cdot$	$\cdot$	$\cdots$	$\cdot$	$\cdot$
v_m	a_{m1}	a_{m2}	$\cdots$	a_{mn}	b_m
	c_1	c_2	$\cdots$	c_n	

Figure: 5.13

We now make two important observations. First, the dual problem to a maximum problem with data box as in Figure 5.12 can be obtained by merely labelling the rows $v_1, \ldots, v_m$; and the dual problem to the minimum problem whose data box is in Figure 5.13 can be obtained by labelling the columns $x_1, \ldots, x_n$. Second, the dimensions of the dual variables in either case can be obtained by dividing the dimensions of the b's or c's by the corresponding a's, as the following examples will make clear. We shall see that the interpretations of the dual variables are easy, once their physical dimensions are determined.

The reader may wonder why we introduce the dual problem instead of concentrating on the original problem alone. The reason is that the simplex method to be discussed later automatically produces the optimum solution to both problems simultaneously. Also, the solution to the dual problem often has important managerial and economic interpretations.

Before we can describe how the simplex method works, we must make a change in the formulation of the dual programmes. What we shall do is to add *slack* variables to the inequalities stated in expressions (1) and (2) of this section in such a way as to make them into equations. To see how this is done, consider as an example the system of inequalities

$- u + 2v \leq 5$, where $u \geq 0$ and $v \geq 0$.

We now add a new slack variable w and obtain a new system of expressions:

$-u + 2v + w = 5$, where $u \geq 0$, $v \geq 0$, $w \geq 0$.

Thus, we obtain the equation

$$-u + 2v + w = 5$$

in non-negative variables. Notice that the new system of expressions is equivalent to the old system, since any solution of the new system that has $w = 0$ represents a case for which $-u + 2v = 5$, and a solution of the new system for which $w > 0$ represents a case for which $-u + 2v < 5$. Moreover, we can write any solution of the old system as a solution of the new system by properly choosing a non-negative value of w. Thus, the truth sets of the two systems are identical.

Now we want to reformulate the constraints of problems (1) and (2). Let y be an m-component vector *of slack variables* y_i, and let f be a number; then (1) is equivalent to

$$\text{Maximise} \quad cx = f$$
$$\text{subject to } Ax + y = b, \qquad (3)$$
$$x, y \geq 0.$$

To see this, rewrite the constraint of (3) as follows:

$$Ax - b = -y; \qquad (4)$$

then $y \geq 0$ is equivalent to $-y \leq 0$, and the latter is, from (4), the same as $Ax \leq b$. The number $f = cx$ measures the current value of the objective function of the maximum problem.

Similarly, let u be an n-component row vector of *slack* variables u_j, and let g be a number; then (2) is equivalent to

$$\text{Minimise} \quad vb = g$$
$$\text{subject to } vA - u = c, \qquad (5)$$
$$u, v \geq 0.$$

To see the equivalence rewrite the equality constraint of (5) as

$$vA - c = u; \qquad (6)$$

then it is obvious that $u \geq 0$ and $vA \geq c$ are the same. The number $g = vb$ measures the current value of the objective function of the minimising problem.

Next we show that the pair of dual problems in (3) and (5) can both be represented in the same tableau, and that the tableau can be obtained by extending either of the data boxes in Figure 5.12 or 5.13. Consider the (Tucker) tableau, which we shall later call the *initial simplex tableau,* in Figure 5.14.

	x_1	x_2	...	x_n	-1	
v_1	a_{11}	a_{12}	...	a_{1n}	b_1	$= -y_1$
v_2	a_{21}	a_{22}	...	a_{2n}	b_2	$= -y_2$
.	.	.	...	.	.	.
v_m	a_{m1}	a_{m2}	...	a_{mn}	b_m	$= -y_m$
-1	c_1	c_2	...	c_n	0	$= f$
	$= u_1$	$= u_2$	...	$= u_n$	$= g$	

Figure: 5.14

Notice that Figure 5.14 can be obtained from Figure 5.12 by adding the 0 entry in the lower right-hand corner, putting variables $v_1, \ldots, v_m$ and -1 along the left margin, putting –1 above the $(n + 1)$st column, marking the right-hand side with $= -y_1, \ldots, = -y_m$, and $= f$, and marking the bottom of the matrix with $= u_1, \ldots, = u_n$, and $= g$. Figure 5.14 can be obtained in a similar manner from Figure 5.13. The reason for this labelling is as follows: if we drop the x's and –1 down to the first row of the matrix, multiply by the coefficients there, and set equal to the label on the right, we have

$$a_{11} + x_1 + a_{12}x_2 + \ldots + a_{1n}x_n - b_1 = -y_1,$$

which is just exactly the first equation of (4). Dropping the labels at the top down to the other rows will give the other equations of (4). Finally, dropping the labels down to the last row gives

$$c_1x_1 + c_2x_2 + \ldots + c_nx_n = f,$$

which is just the definition of f.

In a similar manner, if we move the labels on the left of Figure 5.14 into each column of the tableau, multiply, and set equal to the label at the bottom, we have the various equations of (6) together with the definition of g.

Example 1: The data box for the automobile/truck example of the preceding section is shown in Figures 5.6 and 5.7; hence its initial simplex tableau is as given in Figure 5.15.

	x_1	x_2	-1	
v_1	(5)	2	180	$= -y_1$
v_2	3	3	135	$= -y_2$
-1	300	200	0	$= f$
	$= u_1$	$= u_2$	$= g$	

Figure: 5.15

The primal equations for this problem corresponding to (4) are

$$5x_1 + 2x_2 - 180 = -y_1,$$
$$3x_1 + 3x_2 - 135 = -y_2,$$
$$300x_1 + 200x_2 = f.$$

The dual equations for this problem corresponding to (6) are

$$5v_1 + 2v_2 - 300 = -u_1,$$
$$3v_1 + 3v_2 - 200 = -u_2,$$
$$180v_1 + 135v_2 \quad = g.$$

These are obtained in the manner described above.

Example 2: The data box for the mining example of the preceding section is shown in Figures 5.9 and 5.10; hence its initial simplex tableau is as given in Figure 5.16.

	x_1	x_2	x_3	-1	
v_1	(6)	2	4	200	$=-y_1$
v_2	2	2	12	160	$=-y_2$
-1	12	8	24	0	$=f$
	$=u_1$	$=u_2$	$=u_3$	$=g$	

Figure 5.16

The primal equations for this problem corresponding to (6) are

$$\begin{aligned} 6v_1+2v_2-12&=u_1, \\ 2v_1+2v_2-8&=u_2, \\ 4v_1+12v_2-24&=u_3, \\ 200v_1+160v_2 \quad &=g, \end{aligned}$$

and the dual equations corresponding to (4) are

$$\begin{aligned} 6x_1+2x_2+4x_3-200&=-y_1, \\ 2x_1+2x_2+12x_3-160&=-y_2, \\ 12x_1+8x_2+24x_3&=f. \end{aligned}$$

We next show that from equations (4) and (6) we can immediately derive *Tucker's duality equation:*

$$g-f=vy+ux. \tag{7}$$

This follows easily since

$$g-f=vb-cx=v(Ax+y)-(vA-u)x=vy+ux,$$

where we used the substitutions $b=Ax+y$ from (4) and $c=vA-u$ from (6).

Non-negative vectors x, y, u, and v that satisfy (4) and (6) will be called *feasible vectors* for the equality form of the linear programming problem. Note that the duality relation (7) is true for *all* solutions x, y, u, and v satisfying (4) and (6) whether non-negative or not. However, the following theorem shows that a pair of feasible vectors for one of the problems implies a bound on the objective function of the other problem.

Theorem: (a) Let x^0, y^0, and f^0 be *optimal* solutions to maximising problem (3), and let u, v, and g be *feasible* solutions to the dual minimising problem (5); then $cx^0 = f^0 \leq g = vb$; in other words, for any feasible vector v, the value g — vb is an *upper bound* to the maximum value $f^0 = cx^0$ of the maximising problem (3).

(b) Let u^0, v^0, and g^0 be *optimal* solutions to the minimising problem (5), and let x, y, and f be *feasible* solutions to the dual maximising problem (3); then $v^0b = g^0 \geq f = cx$; in other words, for any feasible vector x, the value $x = cx$ is a *lower bound* to the minimum value $g^0 = v^0b$ of the minimising problem (5).

Proof: (a) If u, v, x^0, and y^0 are all non-negative vectors, then it follows that $vy^0 \geq 0$ and $ux^0 \geq 0$, so that, from (7), we have

$$g - f^0 = vy^0 + ux^0 \geq 0,$$

or, in other words, $g \geq f^0$, as asserted.

The proof of (b) is similar.

We illustrate the theorem by returning to the previous examples.

Example 1: If we consider the automobile/truck example whose initial tableau is given in Figure 5.15, we can easily check that the following quantities solve the primal problem: $x_1 = 10$, $x_2 = 10$, $y_1 = 110$, $y_2 = 75$. These were obtained by selecting arbitrary but not too large values for x_1 and x_2 and then solving for y_1 and y_2. From this feasible solution we calculate $cx = 300 \cdot 10 + 200 \cdot 10 = 3000 + 2000 = 5000$; hence we know that $5000 \leq g^0 = v^0b$; that is, we have found a lower bound to the optimum value g^0 of the dual minimising problem.

Similarly, we can select v_1 and v_2 to be fairly large, but otherwise arbitrary, and solve for u_1 and u_2. For instance, $v_1 = 50$, $v_2 = 40$, $u_1 = 70$, and $u_2 = 20$ are a feasible choice for these quantities. From them we know that $f^0 = cx^0$ is definitely not greater than $vb = 180 \cdot 50 + 135 \cdot 40 = 9000 + 5400 = 14{,}400$.

Since we know that the optimum value is $f^0 = 12{,}000$, and we will later show that f^0 — g^0, we see that, in fact, the lower and upper bounds are correct in this instance. The reader should try several other feasible solutions for this example.

Example 2: Let us check the theorem for the mining example shown in Figure 5.16. Suppose we choose $x_1 = 20$, $x_2 = 20$, $x_3 = 5$, so that

$y_1 = 20$ *and* $y_2 = 20$. We thus obtain the lower bound on g^0 as $cx = 12 \cdot 20 + 8 \cdot 20 + 24 \cdot 5 = 240 + 160 + 120 = 520$.

Similarly, we can choose $v_1 = 2$, $v_2 = 2$, and correspondingly $u_1 = 4$, $u_2 = 0$, and $u_3 = 8$, so an upper bound for f^0 is $vb = 200 \cdot 2 + 160 \cdot 2 = 720$.

Since the true value is 680, we see that the upper and lower bounds again are correct.

Exercises

1. Illustrate the theorem of this section by finding other feasible solutions to the primal and dual problems for the automobile/truck example, and show that the upper and lower bounds so obtained are correct.
2. Repeat Exercise 1 for the mining example.
3. Let x^0 and v^0 be non-negative vectors such that $f^0 = cx^0 = v^0 b = g^0$, $Ax^0 \le b$, and $v^0 A \ge c$.

   (a) Show that if x is any other feasible vector, then

   $cx \le v^0 Ax \le v^0 b = cx^0$,

   so that x^0 solves the maximum problem.

   (b) Similarly, show that v^0 solves the minimum problem.

   (c) Show that $cx^0 = v^0 b = v^0 Ax^0$.
4. Use (7) to show that if x, y, u and v are vectors related as in (4) and (6), then $ux \ge 0$ and $vy \ge 0$ imply $g \ge f$. (Note that this is true whether or not x, y, u, and v are non-negative.)

   If x, y, u, and v are vectors related as in (4) and (6), then they are said to have the *complementary slackness property* if and only if

   $$ux = 0 \text{ and } vy = 0.$$

The remaining exercises refer to this property.

5. Use (7) to show that if x, y, u, and v satisfy the complementary slackness property, then $g = f$. Is the converse true?
6. If x, y, u, and v are non-negative vectors, show that $g = f$ if and only if they have the complementary slackness property.
7. Use Exercises 3, 5, and 6 to show that non-negative vectors related as in (4) and (6) are optimal if and only if they satisfy the complementary slackness property.

Simplex Method

Earlier we solved simple linear programming problems having two variables by sketching convex sets in the plane. To solve such problems in more than two variables by the same method would require visualising convex sets in more than two dimensions, which is extremely difficult. But fortunately there is an algorithm, called the *simplex algorithm,* that permits us to solve such large-scale linear programming problems without such visualisation. The reader will recall that earlier we developed an algorithm for solving simultaneous linear equations that was algebraic (not geometric) in nature and avoided similar visualisation problems.

For simplicity we shall make the following two assumptions in the present and next sections:

1. ***The Non-negativity Assumption:*** We shall assume $b \geq 0$; that is, every component of b is non-negative.
2. ***The Non-degeneracy Assumption:*** The extreme points of the convex set of feasible vectors are each the intersection of *exactly n* bounding hyperplanes, where n is the number of components of the vectors involved.

We shall indicate how these two assumptions can be dropped. We emphasise, however, that for linear programming problems derived from actual applications both assumptions will be satisfied, or else the problem can be reformulated so that they are. Moreover, when codes are written for computers to solve linear programming problems, precautions are taken to insure that these assumptions hold.

We now proceed to describe the simplex method. In the next section we shall discuss reasons why the simplex method works.

After the data box has been set up for either a maximising or minimising problem, the simplex method begins with the initial simplex tableau (the Tucker tableau) of Figure 5.14. Note that it was derived from the data box as described in the previous section. The simplex algorithm will change the initial tableau into a second one, that into a third, and so on, until finally a tableau is obtained that displays the optimum answers to both the primal and dual problems. A typical tableau in this computational process is shown in Figure 5.17. Note that the variables have been identified as being of two kinds: *basic* and *non-basic.*

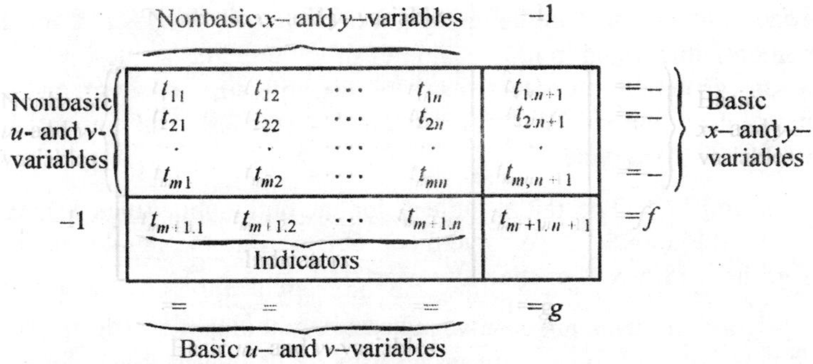

Figure: 5.17

The basic variables appear on the bottom and right-hand sides of the tableau and the non-basic variables on the left and top. As we shall see, in any tableau, if we set the non-basic variables equal to zero, then the corresponding values of the basic variables can be read from the last row and last column of the tableau. The other important thing to note is that the entries of the first *n* columns of the last row are called *indicators.*

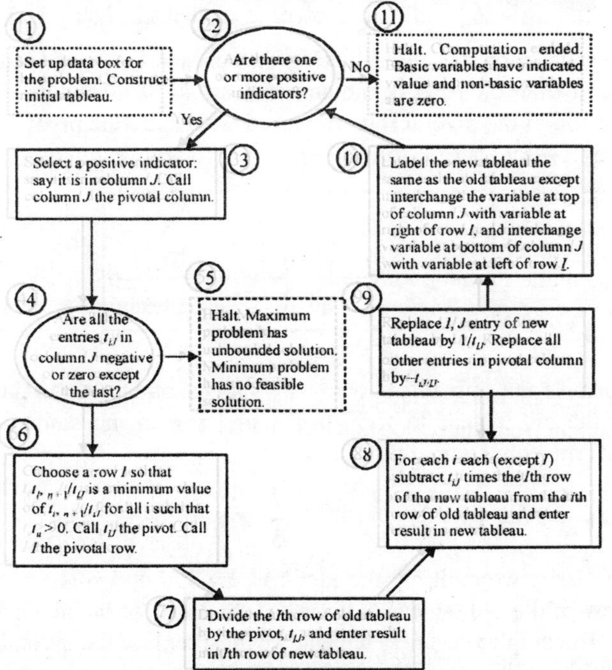

Figure: 5.18

The flow chart in Figure 5.18 describes how the simplex method works. Look at box 1 in the upper left-hand corner. We see that for the automobile/truck and mining examples of the previous section we have already carried out the directives there: the problems are set up and the initial tableaus formed. We now discuss in detail the rest of the computation for these two examples.

Example 1: The initial tableau for the automobile/truck example appears in Figure 5.15. To solve this problem using the simplex method we go next to box 2 of the flow chart in Figure 5.18.

Figure 5.15 there are positive indicators, so the answer to the question in box 2 is "yes." Hence, we proceed to box 3, which says, "Select a positive indicator." Suppose we select 300, which makes column 1 the pivotal column and $J = 1$. We now go to box 4 and observe that there are positive entries in column 1, so that the answer to the question there is "no," and we go on to box 6. We must now find the pivotal row. For this we examine the ratios $t_{i,n+1}/t_{i1}$ for $i = 1$ and 2. These ratios are $180/5 = 36$ and $135/3 = 45$. Since the smaller ratio occurs in the first row, we see that the 5 entry in the first column of Figure 5.15 is the pivot and $I = 1$, so that the first row is pivotal. The pivot is circled in Figure 5.15.

Next we carry out the directives in boxes 7 and 8 of Figure 5.18, which construct the rows of the new tableau. In box 7 we find we must divide through the pivotal row of the old tableau by the pivot and insert it in the new tableau (Figure 5.19).

1	$\frac{2}{5}$	36
0	$\frac{9}{5}$	27
0	80	−10,800

Figure: 5.19

Then we multiply this new row by 3 and subtract it from the second row of the old tableau to form the second row of the new tableau. In vector form, this computation is

$$-3(1 \quad \frac{2}{5} \quad 36) + (3 \quad 3 \quad 135) = (0 \quad \frac{9}{5} \quad 27).$$

Similarly, we multiply the new row by 300 and subtract from the third row of the old tableau to form the third row of the new tableau as shown. To complete the new tableau we must replace the pivotal column as described in box 9 of Figure 5.18; the result is given in Figure 5.20.

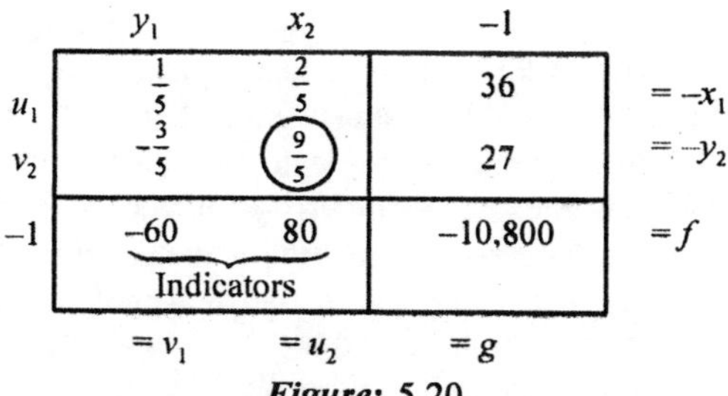

	y_1	x_2	-1	
u_1	$\frac{1}{5}$	$\frac{2}{5}$	36	$= -x_1$
v_2	$-\frac{3}{5}$	$\frac{9}{5}$ (circled)	27	$= -y_2$
-1	-60	80	$-10{,}800$	$= f$
	Indicators			
	$= v_1$	$= u_2$	$= g$	

Figure: 5.20

Also we must interchange the labels of the variables at both ends of the pivot row with the variables at both ends of the pivot column as described in box 10 of Figure 5.18. The completed new second tableau appears in Figure 5.20.

We now find ourselves back at box 2 of the flow chart of Figure 5.18. Since the 80 in the second column, last row of Figure 5.20 is positive, the answer to the question in box 2 is "yes," so we go to box 3. Clearly we must choose $J = 2$. The answer to question in box 4 is "no," so we go on to box 6 to choose the pivot. The two ratios to be considered are $36/\frac{2}{5} = 90$ and $27/\frac{9}{5} = 15$, so that the second row is pivotal and $\frac{9}{5}$ (circled in Figure 5.20) is the new pivot. Carrying out the instructions in boxes 7 and 8 of the flow diagram gives the tableau in Figure 5.21, and finishing up with boxes 9 and 10 gives the completed third tableau (Figure 5.22).

$\frac{1}{3}$	0	30
$-\frac{1}{3}$	1	15
$-\frac{100}{3}$	0	$-12{,}000$

Figure: 5.21

	y_1	y_2	-1	
u_1	$\frac{1}{3}$	$-\frac{2}{9}$	30	$= -x_1$
u_2	$-\frac{1}{3}$	$\frac{5}{9}$	15	$= -x_2$
-1	$-\frac{100}{3}$	$-\frac{400}{9}$	$-12{,}000$	$= f$
	Indicators			
	$= v_1$	$= v_2$	$= g$	

Figure: 5.22

We again find ourselves in box 2 of the flow diagram. But this time we find no positive indicators for the tableau of Figure 5.22; hence the answer to the question there is "no" and we go to box 11, which says that the computation is ended. The answers to both the primal and dual problems are displayed in the final tableau. To see what they are, we first set the non-basic variables equal to zero as instructed in box 11 of the flow diagram. Hence, we have $u_1 = u_2 = y_1 = y_2 = 0$, since the non-basic variables appear on the left and top of the final tableau. Knowing that $y_i = 0$ for $i = 1, 2$, we drop the variables at the top of the final tableau down to the first row and multiply, obtaining $-30 = -x_1$ or simply $x_1 = 30$. Dropping these down one row further gives $x_2 = 15$. And dropping them down to the last row gives $f = 12{,}000$, which is the final value of the objective function. Thus, the optimal solution vectors to the maximising problem are:

$$x^0 = \begin{pmatrix} 30 \\ 15 \end{pmatrix}, \; y^0 = \begin{pmatrix} 0 \\ 0 \end{pmatrix}, \text{ and } f^0 = 12{,}000.$$

Note that this is the same solution that we found in the previous section. We can also find the optimal solution to the dual problem. (The interpretation of this solution will be given in the next section.) Knowing $U_j = 0$ for $j = 1, 2$, we move the variables on the left of Figure 5.22 into the first column, multiply, and obtain $v_1 = \frac{100}{3}$. Moving them to the second column gives $v_2 = \frac{400}{9}$, and moving them to the third column gives $g = 12{,}000$, the value of the objective function of the minimising problem. Hence, the optimal solution vectors to the minimising problem are:

$$v^0 = \left(\frac{100}{3}, \frac{400}{3}\right), u^0 = (0,0) \text{ and } g^0 = 12{,}000.$$

The reader should substitute x^0 and y^0 into the primal, and v^0 and u^0 into the dual equations written down previously and show that they are satisfied. Note also that $f^0 = v^0 b = cx^0 = g^0$ at an optimum solution. This is always true, and will be discussed further in the next section.

Example 2: Let us solve the mining example using the simplex method. The initial tableau is in Figure 5.16. The first indicator 12 was selected so that the first column is pivotal. The pivot is 6, which is circled in the figure, and was chosen because the two ratios involved are $\frac{100}{3}$ which is smaller than $\frac{160}{2} = 80$, hence the first row is pivotal and the pivot is

6. Carrying out steps 7 through 10 of the flow diagram (Figure 5.18), we construct the second tableau in Figure 5.23.

	y_1	x_2	x_3	-1	
u_1	$\frac{1}{6}$	$\frac{1}{3}$	$\frac{2}{3}$	$\frac{100}{3}$	$= -x_1$
v_2	$-\frac{1}{3}$	$\boxed{\frac{4}{3}}$	$\frac{32}{3}$	$\frac{280}{3}$	$= -y_2$
-1	-2	4	16	-400	$= f$
	Indicators				
	$= v_1$	$= u_2$	$= u_3$	$= g$	

Figure: 5.23

There are two positive indicators, and we choose the first one, 4, so that the second column is pivotal. The new pivot is $\frac{4}{3}$, which is circled in the second (pivotal) row. Carrying out the rest of the steps of the flow diagram, we obtain the third tableau (Figure 5.24).

	y_1	y_2	x_3	-1	
u_1	$\frac{1}{4}$	$-\frac{1}{4}$	-2	10	$= -x_1$
u_2	$-\frac{1}{4}$	$\frac{3}{4}$	8	70	$= -x_2$
-1	-1	-3	-16	-680	$= f$
	Indicators				
	$= v_1$	$= v_2$	$= u_3$	$= g$	

Figure: 5.24

All indicators in this tableau are negative, so the computation is complete. We read off the optimum answers to the primal minimising problem as

$$v^0 = (1, 3),\ u^0 = (0, 0, 16), \text{ and } g^0 = 680,$$

and the final minimum operating cost for the mines is \$680 per week. These are the same answers as the graphical procedure gave.

The optimum answers to the dual maximising problem can also be obtained as

$$x^0 = \begin{pmatrix} 10 \\ 70 \\ 0 \end{pmatrix},\ y^0 = \begin{pmatrix} 0 \\ 0 \end{pmatrix}, \text{ and } f^0 = 680.$$

Example 3: Our next example illustrates the fact that a given variable may first be basic, become non-basic, then become basic again, and so on, several times during the course of the simplex computation. Figures 5.25 through 5.28 give the necessary tableaus, and the pivots are circled there. There is another way of working this problem that requires only two tableaus. It starts with a pivot in the first column instead of the second. This illustrates the rule that it is frequently (but not invariably) better to start the simplex method with a column having the most positive indicator. Note that y_1 started out basic, became non-basic, then became basic again.

	x_1	x_2	-1	
v_1	2	(1)	3	$= -y_1$
v_2	3	1	4	$= -y_2$
-1	17	5	0	$= f$
	$= u_1$	$= u_2$	$= g$	

Figure: 5.25

	x_1	y_2	-1	
u_2	2	1	3	$= -x_2$
v_2	(1)	-1	1	$= -y_2$
-1	7	-5	-15	$= f$
	$= u_1$	$= v_1$	$= g$	

Figure: 5.26

	y_2	y_1	-1	
u_2	-2	(3)	1	$= -x_2$
u_1	1	-1	1	$= -x_1$
-1	-7	2	22	$= f$
	$= v_2$	$= v_1$	$= g$	

Figure: 5.27

	y_2	x_2	-1	
u_1	$-\frac{2}{3}$	$\frac{1}{3}$	$\frac{1}{3}$	$= -y_1$
u_1	$\frac{1}{3}$	$\frac{1}{3}$	$\frac{4}{3}$	$= -x_1$
-1	$-\frac{17}{3}$	$-\frac{2}{3}$	$-22\frac{2}{3}$	$= f$
	$= v_2$	$= u_2$	$= g$	

Figure: 5.28

And x_2 was initially non-basic, became basic, and ended up non-basic. The final optimal answers are:

$$v^0 = \left(0, \frac{17}{3}\right), \qquad u^0 = \left(0, \frac{2}{3}\right), \qquad g^0 = 22\frac{2}{3},$$

$$x^0 = \begin{pmatrix} \frac{4}{3} \\ 0 \end{pmatrix}, \qquad y0 = \begin{pmatrix} \frac{1}{3} \\ 0 \end{pmatrix}, \qquad f^0 = 22\frac{2}{3}.$$

Example 4: The reader has undoubtedly wondered about box 5 of the flow diagram in Figure 5.18, since we have not yet ended in it. Actually, if we are solving an applied problem that is correctly formulated so that it has a solution, we shall never end in it. Consider, however, the problem whose initial tableau is in Figure 5.29. Both the first two columns have positive indicators. If we choose the first one and pivot, we obtain the tableau of Figure 5.30. Now there is one positive indicator in the second column, so J = 2.

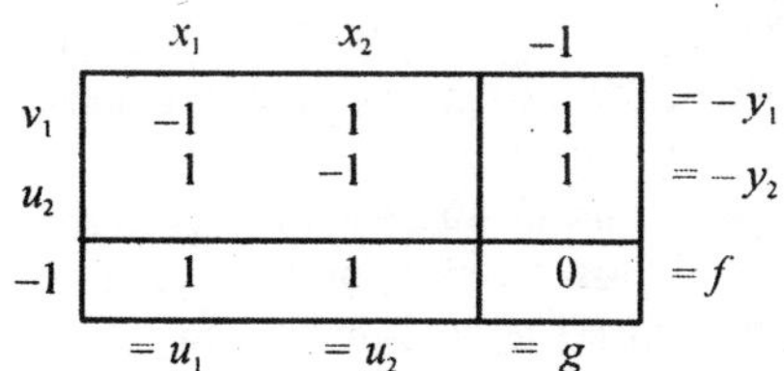

	x_1	x_2	-1	
v_1	-1	1	1	$= -y_1$
u_2	1	-1	1	$= -y_2$
-1	1	1	0	$= f$
	$= u_1$	$= u_2$	$= g$	

Figure: 5.29

	y_2	x_2	-1	
v_1	1	0	2	$= -y_1$
u_2	1	-1	1	$= -x_1$
-1	-1	2	-1	$= g$
	$= v_2$	$= u_2$	$= g$	

Figure: 5.30

But the answer to the question in box 4 of Figure 5.18 is "yes," so we arrive at box 5, which says that the maximum problem has an unbounded solution and the minimum problem has no feasible solution.

To see this let us write the constraints for the maximum problem of Figure 5.29. They are

$$-x_1 + x_2 \le 1, \qquad x_1 \ge 0,$$

$$x_1 - x_2 \le 1, \qquad x_2 \ge 0.$$

These inequalities are satisfied if x_1 and x_2 are equal and positive. Hence, we can make the objective function $f = x_x + x_2$ as large as we wish. Two constraints of the minimum problem of Figure 5.29 are

$$-v_1 + v_2 \geq 1,$$
$$v_1 - v_2 \geq 1$$

If we add these, we obtain the contradiction 0 > 2, and hence the minimum problem has no solution.

For practical purposes, however, we can ignore the no-solution possibility, since we will be dealing with well-formulated problems that have solutions.

Exercises

1. Solve the problem in Example 3 by choosing the first pivot in the first column. Show that the answer can be obtained in one step.
2. A nut packager has on hand 121 pounds of peanuts and 49 pounds of cashews. He can sell two kinds of mixtures of these nuts: a cheap mix that has 80 per cent peanuts and 20 per cent cashews, or a party mix that has 30 per cent peanuts and 70 per cent cashews. If he can sell the party mix at 80 cents a pound and the cheap mix at 50 cents a pound, how many pounds of each mix should he make in order to maximise the amount he can obtain? *[Ans.* Let x_l be the number of pounds of party mix and x_2 the number of pounds of the cheap mix. Then the data are

   $$A = \begin{pmatrix} .3 & .8 \\ .7 & .2 \end{pmatrix}, \quad b = \begin{pmatrix} 121 \\ 49 \end{pmatrix}, \text{ and } c = (80{,}50).$$

   The packager should make 30 pounds of the party mix and 140 pounds of the cheap mix. His income is $94.]
3. The operator of all oil refinery can buy light crude oil at $6 per barrel and heavy crude at $5 per barrel. The refining process produces the following quantities of gasoline, kerosene, and fuel oil from one barrel of each type of crude:

Type	*Gasoline*	*Kerosene*	*Fuel Oil*
Light crude	.5	.25	.2
Heavy crude	.4	.3	.25

Note that in each case 5 per cent of the barrel of crude is lost in the form of gases (which have to be burned) and unusable sludge. During the summer months the operator has contracted to deliver 50,000 barrels of gasoline, 30,000 barrels of kerosene, and 10,000 barrels of fuel oil per month. How many barrels of each type of crude should he process in order to meet his production quotas at minimum possible cost?

4. During the winter months the refinery operator of Exercise 12 contracts to deliver 36,000 barrels of gasoline, 12,000 barrels of kerosene, and 18,000 barrels of fuel oil. What is his optimal winter production plan?
5. In Exercises 3 and 4 show that there is an excess production of at least one of the goods during each time of the year. Discuss practical ways in which this excess production can be used.
6. In the tableau of Figure 5.16 make the pivot be the 2 entry in the first column rather than the circled 6 entry shown. Show that this leads to a negative value of x_1 and hence explain the reasons in box 6 of Figure 5.18 for the special choice of the pivot.

Duality Interpretations and Results

As we saw earlier, the Simplex Method is the same for both maximising and minimising problems. The only difference in setting up the two problems is the choice of row or column vectors for the various quantities involved. In either case, we ended up with a data box containing a matrix *A,* a column vector *b,* and a row vector *c.* Using these data, we stated both a maximising and a minimising problem—only one of which initially interested us. The other problem is called the *dual* Linear Programming problem. The dual of a maximising problem is a minimising problem, and viçe versa. And the dual of the Dual Problem is, in either case, the original problem.

We saw that the Simplex Method solves both the original problem and its dual simultaneously. It is, therefore, of interest to see what interpretation, if any, can be given to the dual of a Linear Programming problem. We shall see that we can always give the Dual Problem mathematical and economic or managerial interpretations that are of considerable interest.

The first step in interpreting the solution to the Dual Problem is that of determining the dimensions of the variables involved. Recall that we set up for each linear programming problem a data box, and the numbers in the data box had dimensions. We now need to determine the dimensions of the variables of both the primal and dual problems. The following rule tells how to do this.

Rule for Determining Dimensions of Variables

(a) The dimension of x_j is the ratio of the dimension of b_i divided by the dimension of a_{ij} for any i.

(b) The dimension of v_i is the ratio of the dimension of c_j divided by the dimension of a_{ij} for any j.

In working with dimensions, we use the rules of ordinary algebra for cancelling and so on, as explained earlier.

Example 1: Let us return to the auto/truck example; its data box is given in Figure 5.6. We have already found the dimensions of the primal variables x_1 (trucks/week) and x_2 (autos/week). Let us use rule (b) above to determine the dimensions of the dual variables v_1 and v_2. For v_1 we have:

$$\text{dimension of } v_1 = (\text{dimension of } c_1)/(\text{dimension } a_{11})$$

$$= \frac{\$}{\text{truck}} \Big/ \frac{\text{Sl- manhr}}{\text{truck}}$$

$$= \frac{\$}{\text{truck}} - \frac{\text{truck}}{\text{S1 - manhr}}$$

$$= \frac{\$}{\text{S1 - manhr}}.$$

In Exercise 1, the reader is asked to show that we would have obtained the same result if we had divided the dimension of c_2 by the dimension of a_{12}. In the same manner, we have

$$\text{dimension of } v_2 = (\text{dimension of } c_1)/(\text{dimension } a_{21})$$

$$= \frac{\$}{\text{truck}} \cdot \frac{\text{truck}}{\text{S2 - manhr}}$$

$$= \frac{\$}{\text{S2 - manhr}}.$$

Figure 5.31 summarises the complete data box for the auto/truck example, indicating the dimensions of all variables and constants.

	$x_1 \frac{\text{trucks}}{\text{week}}$	$x_2 \frac{\text{autos}}{\text{week}}$	Capacities
$v_1 \frac{\$}{\text{S1-manhr}}$	$5 \frac{\text{S1-manhr}}{\text{truck}}$	$2 \frac{\text{S1-manhr}}{\text{auto}}$	$180 \frac{\text{S1-manhr}}{\text{week}}$
$v_2 \frac{\$}{\text{S2-manhr}}$	$3 \frac{\text{S2-manhr}}{\text{truck}}$	$3 \frac{\text{S2-manhr}}{\text{auto}}$	$135 \frac{\text{S2-manhr}}{\text{week}}$
Profits	$300 \frac{\$}{\text{truck}}$	$200 \frac{\$}{\text{auto}}$	

	$x_1 \frac{\$}{\text{tons-HG}}$	$x_2 \frac{\$}{\text{ton-MG}}$	$x_2 \frac{\$}{\text{ton-LG}}$	Costs
$v_1 \frac{\text{M1-days}}{\text{week}}$	$6 \frac{\text{tons-HG}}{\text{M1-day}}$	$2 \frac{\text{tons-MG}}{\text{M1-day}}$	$4 \frac{\text{tons-LG}}{\text{M1-day}}$	$200 \frac{\$}{\text{M1-day}}$
$v_2 \frac{\text{M2-days}}{\text{week}}$	$2 \frac{\text{tons-HG}}{\text{M2-day}}$	$2 \frac{\text{tons-MG}}{\text{M2-day}}$	$12 \frac{\text{tons-LG}}{\text{M2-day}}$	$160 \frac{\$}{\text{M2-day}}$
Requirements	$12 \frac{\text{tons-HG}}{\text{week}}$	$8 \frac{\text{tons-MG}}{\text{week}}$	$24 \frac{\text{tons-LG}}{\text{week}}$	

Figure 5.32

Example 2: The data box for the mining example is given in Figure 5.9. We already know that the dimensions of v_1 and v_2 are mine 1-days/week and mine 2-days/week, respectively. Let us use rule (a) above to find the dimensions of x_1.

dimension of x_1 = (dimension of b_1)/(dimension of a_{11})

$$= \frac{\$}{\text{M1}-\text{day}} \bigg/ \frac{\text{tons-Hg}}{\text{M1}-\text{day}}$$

$$= \frac{\$}{\text{tons-Hg}}$$

A similar application of rule (a) gives the dimensions of x_2 and x_3 as \$/ton-Mg and \$/ton-LG, respectively.

Figure 5.32 shows the data box for the mining example, indicating dimensions for all variables and constants.

Determining the dimensions of the dual variables is the first step in their interpretation. The next step is to look at the optimal dual solutions for the examples above and give their interpretations.

Example 1 (continued): In Example 1, we found the optimal solution to the auto/truck example to be

$$x_0 = \begin{pmatrix} 30 \\ 15 \end{pmatrix}, \qquad v^0 = \left(\tfrac{100}{3}, \tfrac{400}{9}\right), \quad f^0 = g^0 = 12{,}000.$$

We know that $v_1^0 = \frac{100}{3}$ has dimensions \$/Sl-manhr, which sound like a *value* for shop 1 man-hours. We shall show that this is in fact the case. Suppose we increase the number of shop 1 man-hours from 180 to 183. Our problem is then summarised in the data box of Figure 5.33, where the dimensions

	x_1	x_2	Capacities
v_1	5	2	183
v_2	3	3	135
Profits	300	200	

Figure 5.33

are the same as in Figure 5.31 and are therefore omitted. The reader will be asked to show in Exercise 2 that the optimal solution to this problem is

$$x^0 = \begin{pmatrix} 31 \\ 14 \end{pmatrix} \text{and } v^0 = \left(\tfrac{100}{3}, \tfrac{400}{9}\right),$$

with objective value 12,100. Notice that the objective value has increased by 100, which is just three times the dual variable $v_1^0 = \frac{100}{3}$. Hence, we see that $v_1^0 = 33.33$ is the *imputed value* of an additional hour of shop 1 man-hours. It should be remarked right away that the imputed-value interpretation holds over only a limited range of changes in shop 1 man-hours. Hence, we should more proberly say that $v_1^0 = 33.33$ is the *imputed value* of an additional hour in shop 1 *provided the dual solution is not changed by* adding this extra capacity.

Note also that the imputed value is determined independently of the *cost* of providing the extra man-hours in shop 1. In order to provide extra man-hours, it would be necessary to pay workers overtime and rent additional equipment, or else do subcontracting, or the like. What the optimal dual variables tell us is the cost of providing extra hours in shop 1 should not be more than their imputed value, or else it is not optimal to get them.

In Exercise 3, the reader will be asked to show that the optimal dual variable $v_2^0 = \frac{400}{9} = 44.44$, which has dimensions \$ per shop 2 man-hour, is the imputed value of an additional hour in shop 2 provided the optimal dual solution does not change after the extra time is added. As before, it is the maximum amount one should be willing to pay to obtain the extra time.

In Example 2 we found the optimal solutions to the mining example to be

$$v^0 = (1,3), \quad x^0 = \begin{pmatrix} 10 \\ 70 \\ 0 \end{pmatrix}, \text{ and } f^0 = g^0 = 680.$$

We know that $x_1^0 = 10$ has dimensions \$ per ton of high-grade ore, which sounds like the *imputed cost* of producing an additional ton of high-grade ore, and we shall show that this is the case. Suppose we increase the requirements for high-grade ore production from 12 to 16 tons. The new data box is shown in Figure 5.34, the dimensions being the same as in Figure 5.32.

	x_1	x_2	x_3	Costs
v_1	6	2	4	200
v_2	2	2	12	160
Requirements	16	8	24	

Figure: 5.34

In Exercise 4, the reader will be asked to show that the optimal solution to the new problem is

$$v^0 = (2,2), \quad x^0 = \begin{pmatrix} 10 \\ 70 \\ 0 \end{pmatrix}, \text{ and } f^0 = g^0 = 720.$$

Notice that the costs of production have increased from 680 to 720, which is $4 \cdot x_1^0 = 4 \cdot 10 = 40$. Hence, $x_1^0 = 10$ was the per-ton cost of each of the additional 4 units of high-grade ore.

In Exercise 5, you will be asked to show that x_2^0 can be similarly interpreted as the *imputed* or *marginal cost* of producing an additional ton of medium-grade ore, *provided* the additional production does *not* cause a new dual solution to appear.

Now let us look at $x_3^0 = 0$, which has dimension \$ per ton of low-grade ore. What this says is that low-grade ore is free in the sense that producing an additional ton has zero cost. What does this mean? If we look at the slack vector $u^0 = (0, 0, 16)$, we observe that there is an over-production of low-grade ore by 16 tons beyond the requirements. In other words we have already overproduced, so the additional ton will cost zero to produce since it already exists. However, this is true only within limits.

	x_1	x_2	x_3	Costs
v_1	6	2	4	200
v_2	2	2	12	160
Requirements	12	8	56	

Figure 5.35

For suppose we change the requirement for low-grade ore to 56 tons, giving the data box of Figure 5.35. In Exercise 6, the reader will be asked to show that the optimal solution to the problem in Figure 5.35 is

$$v^0 = (.5, 4.5), \quad x^0 = \begin{pmatrix} 27.5 \\ 0 \\ 8.75 \end{pmatrix}, \text{ and } f^0 = g^0 = 820.$$

Note that we now have a new dual solution, so that the old dual variable $x_3 = 0$ did not hold for the entire range of changes in the requirements for low-grade ore.

Let us try to give general interpretations to a pair of dual linear programming problems. For either problem, the matrix A will be called the matrix of technological coefficients, since it indicates how activity vectors are combined into the constraining inequalities. Then we can give different interpretations to the vectors *c, b, x,* and *u,* depending on whether our original problem is a maximising or a minimising one.

If the original problem is maximising, we interpret x as the *activity vector.* Then the vector b is interpreted as the *capacity-constraint vector,* whose components give the amounts of the various "scarce resources" that can be demanded by a given activity vector. The vector c is the *profit* vector, whose entries give the unit profits for each component of the activity vector x. Finally, the vector v is the *imputed-value vector,* whose entries give the imputed values of each of the scarce resources that enter into the production process, provided the changes in scarce resources are sufficiently small that the dual solution remains optimal.

If the original problem is minimising, we interpret v as the *activity vector.* Then c is interpreted as the *requirements vector,* whose components give the minimum amounts of each good that must be produced. The vector b is the *cost vector,* whose entries give the unit costs of each of the activities. Finally, the vector x is the *imputed-cost* vector, whose components give the imputed costs of producing additional amounts of each of the required goods, *provided* the changes in requirements are sufficiently small that the dual solution remains optimal.

Next, we shall briefly discuss two important theorems in linear programming. First, we restate the dual problems:

The MAXIMUM Problem		*The MINIMUM Problem*	
Maximise	$cx = f$	Minimise	$vb = g$
Subject to	$Ax + y = b,$ (1)	*Subject to*	$vA - u = c,$ (3)
	$x \geq 0, y \geq 0.$ (2)		$u \geq 0, u \geq 0.$ (4)

Vectors x and y satisfying (1) and (2) and vectors v and u satisfying (3) and (4) are called *feasible vectors.*

In all the examples solved above, we found that $f = g$ at the optimum solution. It is no accident that the dual problems share common values. The next theorem, which is the principal theorem of linear programming, shows that this will always happen whenever the problems have solutions.

The Duality Theorem: The maximum problem has as a solution a feasible vector x^0, such that $cx^0 = \max cx$, if and only if the minimum problem has a solution that is a feasible vector v^0, such that $v^0b = \min$

vb. Moreover, the equality $cx^0 = v^0b$ holds if and only if x^0 and v^0 are solutions to their respective problems.

The duality theorem is extremely powerful, for it says that if one of the problems has a (finite) solution, then the other one necessarily also has a (finite) solution, and both problems share a common value. Another consequence of the theorem is that if one of the problems does *not* have a solution, then neither does the other.

The proof of the duality theorem is beyond the scope of this book, but some parts of it are indicated in Exercises 25 and 26. We saw an example of a Linear Programming problem without a solution earlier also. Another example is in Exercise 27.

The duality theorem states that $g^0 = f^0$ at the optimum solution. Applying this to Tucker's duality equation, we obtain:

$$0 = g^0 - f^0 = v^0y^0 + u^0x^0. \qquad (5)$$

However, since v^0, y^0, u^0, and x^0 are all feasible optimal vectors, they are, in particular, non-negative. Hence, $v^0y^0 \geq 0$ and $u^0x^0 \geq 0$. But the only way that two non-negative numbers can add up to zero is for both of them to be zero. Therefore,

$$v^0y^0. = 0, \qquad (6)$$

$$u^0x^0 = 0. \qquad (7)$$

If we now simply restate (6) and (7), we obtain the following important theorem:

Complementary Slackness Theorem

(A) For each i, *either* $v_i^0 = 0$ or $y_i^0 = b_i - \sum_{i=1}^{n} a_{ij}x_j^0 = 0.$

(B) For each j, *either* $x_i^0 = 0$ or $u_i^0 = \sum_{i=1}^{m} v_1^0 a_{ij} - c_j = 0.$

Proof: The proof of this theorem is simple because (6) says that the sum of the products $v_i^0 y_i^0$ must equal zero, but each term of the product is non-negative so each product must itself be zero, which gives (A). The proof of (B) follows similarly from (7).

Example 2 (continued): From the final tableau in Figure 5.24 we found that the complete solution to the mining problem to be

$$v^0 = (1,3),\ u^0 = (0,0,16),\ x^0 = \begin{pmatrix} 10 \\ 70 \\ 0 \end{pmatrix},\ y^0 = \begin{pmatrix} 0 \\ 0 \end{pmatrix}.$$

We see that since $u_3^0 = 16$—that is, in the optimal solution low-grade ore is overproduced—the imputed cost of low-grade ore must be zero; and it is, since $x_3^0 = 0$. Also, since both v_1^0 and v_2^0 are positive, both components of y^0 must be zero, which they are. The reader should state the other consequences of the complementary slackness theorem for this example.

Let us conclude by discussing the reasons for the various steps of the simplex method. If we always think of the non-basic variables, which appear at the left and on the top of the tableaus, as being set equal to zero, then in the initial tableau of Figure 5.14 we see the initial solution vectors

$$x = 0,\ y = b,\ v = 0,\ \text{and}\ u = -c. \qquad (8)$$

Since we have assumed $b \geq 0$, we see that the first three vectors are non-negative, but u is non-negative only if c was initially non-positive. In the latter case, the initial tableau is optimal. Since this is not normally the case, there is usually at least one positive indicator, so that the first answer to the question in box 2 of Figure 5.18 is "yes." Thus, we must go around the loop and carry out at least one pivot. As we do so, the simplex method systematically changes the tableau in order to make u into a non-negative vector without destroying the non-negativeness of x, y, or v, and also keeping $f = cx = vb = g$ at all times.

In step 6 of Figure 5.18 the pivot was chosen in order to have the smallest ratio $t_{1,n+1} / t_{IJ}$ so that no current x_i or y_j should become negative. The reader may verify that if the pivot is chosen not to have this property, then some such variable is made negative.

The non-degeneracy assumption can be used to show that on each pivot step the value of the current f will actually increase. In Exercise 26 you will be asked to show that at most a finite number of pivot steps can be made. Hence, if the problem has a solution, we must arrive in a finite number of steps at a tableau having all positive indicators. At each step, the current solution in a tableau satisfies equations (1), (2), and (3) above, and when all indicators are positive we have also satisfied (4), so that $u \geq 0$ and $u \geq 0$. By the duality theorem, if we have found x^0, y^0, v^0 and u^0 satisfying (l)-(4) and also $f^0 = cx^0 = v^0b = g^0$, then an optimum solution to the programming problem has been found.

Exercises

1. In Example 1 show that the same answer for the dimension of v_1 can be obtained by dividing the dimension of c_2 by the dimension of a_{12}.

2. Show that the vectors $x^0 = \begin{pmatrix} 31 \\ 14 \end{pmatrix}$ and $v^0 = \left(\frac{100}{3}, \frac{400}{9}\right)$

   solve the problem in Figure 5.33. *[Hint:* Substitute into the primal and dual problems.]

3. (a) Use the optimal solution to the automobile/truck problem in Figure 5.31 to predict how the objective function, which measures profit, will change if the capacity of shop 2 is changed from 135 to 144 man-hours per week.

   (b) Solve the problem in Figure 5.31 with the 135 changed to 144 and use its solution to show that your prediction in (a) was correct.

   [*Ans.* Profit 12,400, $x^0 = \begin{pmatrix} 28 \\ 20 \end{pmatrix}$, $v^0 = \left(\frac{100}{3}, \frac{400}{9}\right)$.]

4. Show that the solution to the mining example in Figure 5.34 is

   $v = (2, 2)$, x^0 as before, $f = g = 720$.

5. (a) Use the solution to Exercise 4 to predict what will happen in the mining problem if the requirement for medium-grade ore is increased from 8 to 10.

   (b) Solve the mining problem in Figure 5.34 with the 8 replaced by 10 and show that your prediction in (a) was correct.

   [*Ans.* $v^0 = (1.5, 3.5)$, x^0 as before, $f = g = 860$.]

6. Show that the solution to the problem in Figure 5.35 is

$$v^0 = (.5,\ 4,\ 5),\ x^0 = \begin{pmatrix} 27.5 \\ 0 \\ 8.75 \end{pmatrix},\quad f = g = 820.$$

   Interpret the solution.

7. In the automobile/truck example of Figure 5.31, suppose that the manufacturer can subcontract up to 18 of either shop 1 or shop 2 man-hours at \$38 per hour. What is his optimal action? *[Hint:* You can answer this question without solving a linear programming problem.]

8. In the mining example of Figure 5.32 suppose the mining owner can sell 10 more tons of medium-grade ore at \$55 per ton. Should he do so?

9. Consider again the general interpretation of a maximising problem in which x is an activity vector, b the capacity-constraint vector, and c the profit vector. Let v^0 be the optimum dual solution vector. Discuss the following managerial interpretation of the components v_i^0 of v^0. "Additional amounts of scarce resource i should be acquired only if its cost is less than the component v_i^0 that gives the imputed value of an additional (sufficiently small) quantity."

10. Consider again the general interpretation of a minimising problem in which v is the activity vector, c the requirements vector, and b the cost vector. Let x^0 be the optimum dual solution vector. Discuss the following managerial interpretations of the components x_i^0 of x. "Additional amounts of the jth good should be produced only if they can be sold with gross profit at least as large as the component x_i^0, which gives the imputed cost of producing an additional (sufficiently small) quantity."

11. Consider the dual maximum and minimum problems in equality form as expressed above. If $c \leq 0$, prove that the initial solution (8) is optimal. *[Hint:* Use the duality theorem.]

12-20. Carry out the following steps:

(a) Find the dimensions of the dual variables.

(b) Set up the initial tableau with the dimensions of all variables and numbers indicated.

(c) Read the answers to both primal and dual problems from the final tableau.

(d) Interpret the dual solutions for the specific problems in each case.

(e) State the complementary slackness theorem for each problem and interpret.

21-24. Rework Exercises 11-13 using steps (a)-(e) of Exercises 12-20, above.

*25. The assumption of non-degeneracy can be shown to be equivalent to the following: At no time in the pivoting process of the simplex method are any of the entries in the first m rows of the last column of the tableau ever zero. Use this fact to show that on each pivot step, the value of $f = ex$ *increases*.

*26. Show that there are only a finite number of ways that the components of the x- and y-vectors can be used to label the top and right-hand side of the various tableaus during the pivoting process.

Use the result of Exercise 25 to show that no tableau can ever be repeated in the course of solving a non-degenerate problem by the simplex method. Hence, conclude that the simplex method described in Figure 5.18 must stop in a finite number of steps with the optimal solution to the linear programming problem, or else with proof that the problem has no finite solution.

27 Use the flow diagram of Figure 5.18 to show that the problem whose initial tableau is

−1	1	4
2	−4	8
2	3	0

does not have a solution. Verify algebraically and geometrically the statements in box 5 of that flow diagram.

Equality Constraints and General Simplex Method

In this (optional) part we shall discuss the removal of the non-negativity and non-degeneracy assumptions that we imposed on linear programming problems. As stated there, most problems will automatically satisfy these assumptions. If not, they can usually be changed so that they do. We illustrate the latter first.

Example 1: Consider again the automobile/truck example of Figure 5.6. Suppose we add the managerial constraint that at least 20 automobiles should be produced —perhaps because we have orders for them. The inequality that will do this is $x_2 \geq 20$, but it is a $\geq$ inequality instead of a $\leq$ inequality as is required for a maximising problem. Multiplying through by -1 gives $-x_3 \leq -20$. Hence, the maximising problem is

$$\text{Maximise} \qquad 300x_1 + 200x_2 \qquad (1)$$

$$\begin{aligned} \text{subject to} \qquad 5x_1 + 2x_2 &\leq 180, \\ 3x_1 + 3x_2 &\leq 135, \\ -x_2 &\leq -20, \\ x_1, x_2 &\geq 0. \end{aligned}$$

We see that the b vector is

$$b = \begin{pmatrix} 180 \\ 135 \\ -20 \end{pmatrix}, \qquad (2)$$

which does not satisfy the non-negativity assumption. However, let us set up the initial tableau and see what we can do with it. It is shown in Figure 5.36. Notice that in the third row where the – 20 entry is, there is also a – 1.

	x_1	x_2	-1	
v_1	5	2	180	$= -y_1$
u_2	3	3	135	$= -y_2$
u_3	0	(–1)	– 20	$= -y_3$
–1	300	200	0	$= f$
	$= u_1$	$= u_2$	$= g$	

Figure: 5.36

If we were to pivot on the –1, using the usual rules as given in Figure 5.18, we could change the –20 into a +20. Carrying out this pivot operation gives the tableau of Figure 5.37, which *has* a positive *b* vector. Hence, we

	x_1	y_3	-1	
v_1	5	2	140	$= -y_1$
v_2	③	3	75	$= -y_2$
v_3	0	1	20	$= -x_2$
-1	300	200	-4000	$= f$
	$= u_1$	$= v_3$	$= g$	

Figure 5.37

can now proceed in the usual way. Choosing the most positive indicator, which is 300, we determine that the pivot should be the 3 circled in the first column. Carrying out the rest of the pivot steps as in Figure 5.18 gives the tableau in Figure 5.38. Since all indicators there are negative, we have

	y_2	y_3	-1	
v_1	$-\frac{5}{3}$	-3	65	$= -y_1$
u_1	$\frac{1}{3}$	1	25	$= -x_1$
u_2	0	1	20	$= -x_2$
-1	-100	-100	$-11{,}500$	$= f$
	$= v_2$	$= v_3$	$= g$	

Figure 5.38

determined the optimal solution, namely

$$x^0 = \begin{pmatrix} 25 \\ 20 \end{pmatrix}, \; v^0 = (0,100,100), \text{ and } f^0 = g^0 = 11{,}500.$$

In other words, the optimum decision now is to produce 25 trucks and 20 automobiles for a gross profit of $11,500. Notice that the gross profit has gone down, which is not surprising since we are satisfying an additional constraint. Notice also that the dual solution indicates that for each auto-mobile less that we require to be made, an additional $100 profit can be realised. This follows because v_3^0 – $100, indicating that if we increase the right-hand side of the third constraint by 1, that is, change – 20 to –19, then the profit should increase by $100. Notice also that the imputed value of shop 1 man-hours has gone to zero! This is because $y_1 = 65$, indicating that we are not using all of the shop 1 man-hours. Also the imputed value of shop 2 man-hours has jumped from \$44.44 to \$100 per hour, which indicates that shop 2 has become a more important "bottleneck" in the production process.

The previous example shows one way of deriving a problem that has negative *b* vector coefficients—namely, by imposing a $\leq$ constraint with positive right-hand side on the maximising problem. Another way is to impose an equality constraint. For example, consider the equation

$$2x_1 + 5x_2 - 7x_3 = 12. \tag{3}$$

We can replace it by the two inequalities

$$2x_1 + 5x_2 - 7x_3 \leq 12 \text{ and } 2x_1 + 5x_2 - 7x_3 \geq 12, \tag{4}$$

but the second of these is a $\geq$ constraint. We can change it into a $\leq$ constraint by multiplying by a -1, obtaining

$$2x_1 + 5x_2 - 7x_3 \leq 12 \text{ and } -2x_1 - 5x_2 + 7x_3 \leq -12 \tag{5}$$

as a pair of $\leq$ inequalities that are equivalent to the single equality (3). When solving simple problems such as in Example 1 by hand it is usually quite easy to see how to pivot on negative numbers in the tableau in such a way that the problem becomes one having non-negative right-hand sides. However, for large problems, and particularly for computing-machine computation, it is necessary to have a set of rules that will always work, without depending upon the ingenuity of the user. Such an algorithm is presented in Figure 5.39. It is usually called "phase I" of the simplex method, and what it does is to put the tableau in the standard form so that the flow diagram of Figure 5.18 can be applied. We illustrate it with an example.

Example 2: Consider the linear programming problem

$$\begin{aligned} &\text{Maximise} && 2x_1 + x_2 \\ &\text{subject to} && x_1 + x_2 \leq 20, \\ & && x_1 + 2x_2 = 30, \\ & && x_1, x_2 \geq 0. \end{aligned} \tag{6}$$

The set of feasible *x*-vectors is the line segment between the points $\begin{pmatrix} 0 \\ 15 \end{pmatrix}$ and $\begin{pmatrix} 10 \\ 10 \end{pmatrix}$ shown darkened in Figure 5.40. In order to solve

(6) we replace the equality constraint by a pair of inequalities and obtain the problem:

$$\begin{aligned} \text{Maximise} \quad & 2x_1 + x_2 \\ \text{subject to} \quad & x_1 + x_2 \leq 20, \\ & x_1 + 2x_2 \leq 30, \\ & -x_1 - 2x_2 \leq -30, \\ & x_1, x_2 \geq 0. \end{aligned} \tag{7}$$

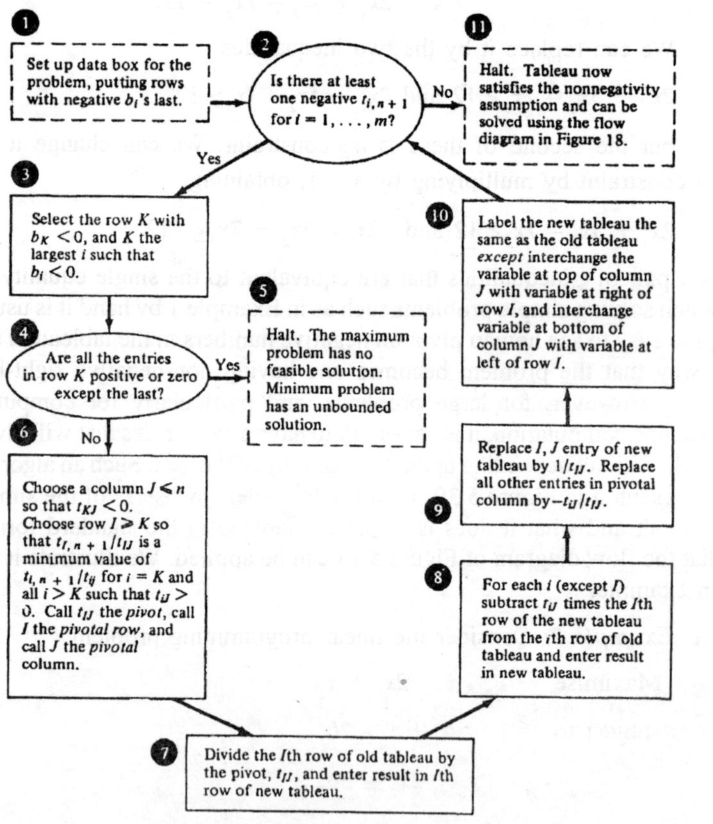

Figure: 5.39

Thus, we obtain a problem that does not satisfy the non-negativity assumption. Let us solve the problem by following the flow diagram of Figure 5.39. We set up the initial tableau with the negative b_i's last as instructed in box 1 of that figure. The initial tableau is given in Figure

5.41. The answer to the question in box 2 of Figure 5.39 is "yes," so we go to box 3, where we must choose $K = 3$. The answer to the question in box 4 is "no," so we go on to box 6. Since both entries in the first two columns of the third row of Figure 5.41 are negative, J can be either 1 or 2; we choose $J=1$. Then the ratio rule in box 6 gives $7 = 3$. Carrying out the pivot steps in boxes 7-10 of Figure 5.39 gives the next tableau shown in Figure 5.42.

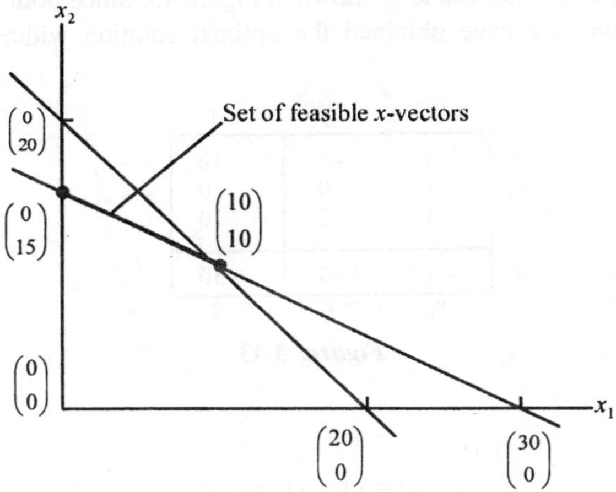

Figure: 5.40

Notice that a new negative has appeared in the third column of the first row! So the answer to the question in box 2 of Figure 5.39 is again "yes," and we must go around the main loop of the flow diagram again. We find that

	x_1	x_2	-1	
v_1	1	1	20	$= -y_1$
v_2	1	2	30	$= -y_2$
v_3	(−1)	−2	− 30	$= -y_3$
−1	2	1	0	$= f$
	$= u_1$	$= u_2$	$= g$	

Figure: 5.41

$K = 1$ and $J = 2$ are the only possible choices, and these give $I = 1$, so that we must pivot on the −1 circled in the first row of Figure 5.42. After

	y_3	x_2	-1	
v_1	1	(−1)	−10	$= -y_1$
v_2	1	0	0	$= -y_2$
u_1	−1	2	30	$= -x_1$
-1	2	−3	− 60	$= f$
	$= u_3$	$= u_2$	$= g$	

Figure: 5.42

pivoting, the new tableau is as shown in Figure 43. Since both indicators are negative, we have obtained the optimal solution without further pivoting.

	y_3	y_1	-1	
u_2	− 1	−1	10	$= -x_2$
v_2	1	0	0	$= -y_2$
u_1	1	2	10	$= -x_1$
-1	− 1	−3	− 30	$= f$
	$= v_2$	$= v_1$	$= g$	

Figure: 5.43

It is

$$x^0 = \begin{pmatrix} 10 \\ 10 \end{pmatrix}, \; V^0 = (3,0,1), \text{ and } f^0 = g^0 = 30.$$

The reader should locate the solution on the diagram of Figure 5.40.

The last topic of this section is the question of removing the non-degeneracy assumption stated earlier. A complete discussion of the problem is beyond the scope of this book, but an interested reader may wish to refer to one of the more advanced texts listed at the end of this chapter. We shall indicate the essential ideas here, however. An example will suffice for this purpose.

Example 3 Consider the problem:

$$\begin{aligned} &\text{Maximise} && x_1 + x_2 \\ &\text{subject to} && x_1 \le 4, \\ &&& x_2 \le 4, \\ &&& 2x_1 + x_2 \le -8, \\ &&& x_1, x_2 \ge 0. \end{aligned} \tag{8}$$

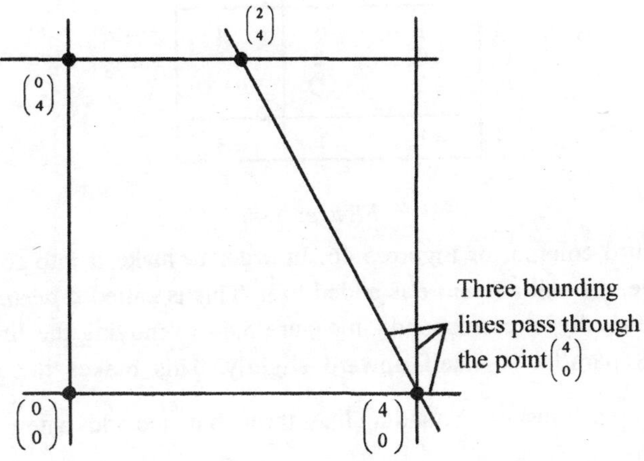

Figure: 5.44

The set of feasible x-vectors is shown shaded in Figure 5.44. Notice that the set has four extreme points and that each is the intersection of exactly two bounding lines *except* for the point $\begin{pmatrix}4\\0\end{pmatrix}$, which has three bounding lines through it. We shall show that this can lead to the appearance of a zero in the b area of the tableau after some pivoting, and when this happens it is possible to pivot without improving the objective function. The initial tableau for the problem is given in Figure 5.45.

	x_1	x_2	-1	
v_1	①	0	4	$= -y_1$
v_2	0	1	4	$= -y_2$
v_3	2	1	8	$= -y_3$
-1	1	1	0	$= g$
	$= u_1$	$= u_2$	$= g$	

Figure: 5.45

Since both indicators are positive, suppose we choose the first one. The minimum-ratio rule then selects the first row to be pivotal, and we pivot on the one circled. (Note that we could also pivot on the 2 in the third row, first column, and the results will be similar; see Exercise 11.) The new tableau is given in Figure 5.46. Notice that a zero did appear in the third

	y_1	x_2	-1	
u_1	1	0	4	$= -x_1$
v_2	0	1	4	$= -y_2$
u_3	-2	①	$0 + \in$	$= -y_3$
-1	-1	1	-4	$= f$
	$= v_1$	$= u_2$	$= g$	

Figure: 5.46

row, third column, of Figure 5.46. In order to make it into something positive, a small amount e is added to it. This is called *a perturbation.* Geometrically, it corresponds in Figure 5.44 to moving the line $2x_1 + x_2 = 8$ parallel to itself upward slightly. This makes the extreme point $\begin{pmatrix} 4 \\ 0 \end{pmatrix}$ have just two bounding lines through it, and adds a new extreme point $\begin{pmatrix} 4 \\ \in \end{pmatrix}$ nearby. We will find it on the next iteration. The second column has a positive indicator, and the minimum-ratio rule selects the third row to be pivotal and 1 the pivot, circled in Figure 5.46. The new tableau is given in Figure 5.47. Now we observe that column 1 has a positive indicator, so we must still

	y_1	y_3	-1	
u_1	1	0	4	$= -x_1$
v_2	②	-1	$4 - \in$	$= -y_2$
u_2	-2	1	$0 + \in$	$= -x_2$
-1	1	-1	$-4 - \in$	$= f$
	$= v_1$	$= v_3$	$= g$	

Figure: 5.47

	y_2	y_3	-1	
u_1	$-\frac{1}{2}$	$-\frac{1}{2}$	$2 + (\frac{\in}{2})$	$= -x_1$
v_2	1	$-\frac{1}{2}$	$2 - (\frac{\in}{2})$	$= -y_1$
u_2	-1	$\frac{1}{\frac{2}{0}}$	4	$= -x_2$
-1	$-\frac{1}{2}$	$-\frac{1}{2}$	$-6 - (\frac{\in}{2})$	$= f$
	$= v_2$	$= v_3$	$= g$	

Figure: 5.48

pivot again. The ratio rule selects as pivot the 2 in the first column, circled in Figure 5.47. The next tableau is given in Figure 5.48.

Since both indicators in Figure 5.48 are negative, we have the optimal solution. Notice that if we replace e by 0 we still have an optimal tableau, hence our perturbation did not affect the original problem enough to change the solution, which is

$$x^0 = \begin{pmatrix} 2 \\ 4 \end{pmatrix}, \ v^0 = \left(0, \tfrac{1}{2}, \tfrac{1}{2}\right), \ \text{and } f^0 = g^0 = 6.$$

Actually, if we had ignored the 0 in the last column of Figure 5.47 and just gone ahead with the Simplex Method as given in Figure 5.18, we would have arrived at the same solution without difficulty. But notice that in going from tableau 46 to 47 we then would not have increased the objective function/at all. It can happen with larger problems that the computation could go from one tableau to the next several times in a row without changing/ and after a finite number of pivots return to a tableau constructed earlier. From then on, the computational process will go through the same sequence of tableaus indefinitely without changing f. This is called *cycling.* Actually, it rarely happens in practice. The smallest known example in which it can occur has seven variables. For small problems that can be worked by hand it never occurs.

There are several ways of avoiding cycling for computer codes that handle large problems. One way is the process of perturbation illustrated above. There only one 0 was found and it was made positive by adding $+\epsilon$ to it. If a second zero were found, then $+\epsilon^2$ would be added; and if a third were found, then $+\epsilon^3$ would be added; and so on. The final tableau will then have numbers plus polynomials in e in the last column. By selecting ϵ *not* to be equal to any of the finite number of zeros of these polynomials and also very small, we can prove that there always is a perturbation of the components of the b-vectors that will avoid cycling, and that has the same solution as the original problem when c is replaced by 0 in the final tableau.

Still another (practical) way of avoiding cycling is the following. Whenever a zero is about to appear in a tableau, there will be more than one choice of pivotal row in box 6 of the flow diagram of Figure 5.18. This can be seen in Figure 5.45, in which, given the pivotal column $J = 1$, we can choose *either* $I = 1$ *or* $I = 3$ when applying the test. Suppose now we choose between these two at random, instead of always choosing the first one. It can be shown that if this method is used to "break ties" when selecting pivotal rows, then the Simplex Method will not cycle with probability 1. For practical purposes, this provides an adequate safeguard against the very rare possibility of cycling in computations.

Exercises

1. Write pairs of $<$ inequalities that are equivalent to each of the following = constraints:

   (a) $12x_1 + 3x_2 - 7x_3 = 15.$

   (b) $3x_1 - 2x_2 + 4x_3 = 0$ and $-4x_1 + x_2 - 2x_3 = 7.$
2. Consider the mining example again with the additional constraint that exactly 16 tons of high-grade ore should be produced per week. Show that the tableau has a non-negative b-vector.
3. Show that a minimising problem with $b \geq 0$ can always be solved using Figure 5.18 regardless of the form of the additional constraints that may be imposed on the minimising problem.
4. Show that the additional constraint $x_1 \leq 15$ can be imposed and the problem solved using Figure 5.18.
5. Show that a maximising problem with only $\leq$ constraints and positive b-vector can be solved using Figure 5.18 regardless of how many additional $\leq$ constraints are added, as long as the right-hand sides of such additional constraints are non-negative.
6. Use the results of Exercises 3-5 to show that the phase I computation of Figure 5.39 is needed only when a $\leq$ constraint with negative right-hand side is added to the maximising problem.
7. Apply the phase I Simplex Method of Figure 5.39 to the following examples:

   (a) Maximise $2x_1 + x_2$

   subject to $x_1 + x_2 < 10,$

   $x_1 + x_2 \geq 6,$

   $x_1 \leq 8$

   $x_1, x_2 \geq 0.$

   (b) Maximise x_1

   subject to $x_1 \geq 2,$

   $x_2 \geq 3,$

   $3x_1 + 2x_2 \leq 24.$
8. Apply the phase I computation to the problem whose initial tableau is given by

1	1	20
− 1	− 2	− 50
2	1	0

and show that the computation ends up in box 5 of Figure 5.39. Draw the constraint sets of the primal and dual problems and give a geometric interpretation to the statements in box 5 of Figure 5.39.

9. Show in general that if the computation of Figure 5.39 ends up in box 5, then the statements given there are correct.
10. Show that phase I is needed if and only if $x = 0$ is *not* a feasible vector for the maximising problem.
11. Start with Figure 5.45 and carry out pivoting steps, starting with the pivot in the third row, first column. Show that equivalent results are obtained.
12. Show that even if we do not add $+\epsilon$ in the third row, third column, of Figure 5.46, the simplex method will yield the correct solution.
13. Add the constraint $-x_1 + x_2 \leq 4$ to the problem in (8) and show that no matter which column is chosen for the first pivot, a 0 is still produced in the b-vector after one pivot. Show that the simplex method still works.
14. (a) Show that the phase I simplex method will eventually make the last inequality with negative right-hand side into one with positive (or zero) right-hand side without making the right-hand sides of later inequalities negative.

    (b) Show that in a finite number of steps all negative right-hand sides will be made non-negative, or else the computation will end up in box 5 of the flow diagram in Figure 5.39.

Strictly Determined Games

We discussed linear programming problems that involve *optimisation*—that is, the maximisation or minimisation of a (lincar) function subject to linear constraints. In order to optimise a function, it is necessary to control all relevant variables.

Game theory considers situations in which there are two (or more) persons, each of whom controls some but not all the variables necessary to determine the outcome(*s*) of a certain event. Depending upon which event actually occurs, the players receive various payments. If for each possible event, the algebraic sum of payments to all players is zero, the game is called *zero-sum;* otherwise, it is *non-zero-sum.* Usually, the players will not agree as to which event should occur, so that their objectives in the game are different. In the case of a matrix game, which is a two-person zero-sum game in which one player loses what the other wins, game theory provides a solution. The solution is based on the principle that each player tries to choose his course of action so that, regardless of what his opponent does, he can assure himself of a certain minimum amount.

Most recreational games such as ticktacktoe, checkers, backgammon, chess, poker, bridge, and other card or board games can be viewed as games of strategy. On the other hand, such gambling games as dice, roulette, and so on are not (as usually formulated) games of strategy, since a person playing one of these games is merely "betting against the odds."

Now, we shall formulate simple games that illustrate the theory and are amenable to computation. We shall base these games on applications in business situations and on recreational games.

Example 1: Two stores, R and C, are planning to locate in one of two towns. As in Figure 5.49, town 1 has 60 per cent of the population while town 2 has 40 per cent. If both stores locate in the same town they will split the total business of both towns equally, but if they locate in different towns each will get the business of that town. Where should each store locate?

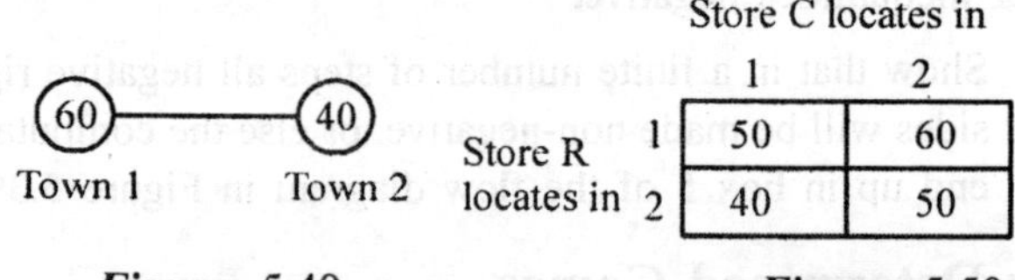

		Store C locates in 1	2
Store R locates in	1	50	60
	2	40	50

Figure: 5.49 ***Figure:*** 5.50

Clearly, this is a game situation, since each store can control where it locates but cannot control at all where its competitor locates. Each store has two possible "strategies": "locate in town 1" and "locate in town 2." Let us list all possible outcomes for each store employing each of its strategies. The result is given in the *pay-off matrix* of Figure 5.50. The entries of the matrix represent the percentages of business that store R gets in each case. They can also be interpreted as the percentage *losses* of business by C for each case. If both stores locate in town 1 or both in town 2, each gets 50 per cent of the business, hence the entries on the main diagonal are 50. If store *R* locates in town 1 and C in 2, then R gets 60 per cent of the business as indicated in entry in row 1 and column 2. (This entry also indicates that C *loses* 60 per cent.) Similarly, if R locates in 2 and C in 1, then R gets 40 per cent (and C loses 40 per cent) as indicated in row 2 and column 1.

How should the players play the matrix game in Figure 5.50? It is easy to see that store R should prefer to locate in town 1 because, regardless of what C does, R can assure himself of 10 per cent more business in town 1 than in town 2. Similarly, store C also prefers to locate in town 1 because he will lose 10 per cent less business—that is, gain 10 per cent more business—in town 1 than in 2. Hence, optimal strategies are for each store to locate in town 1; that is, R chooses row 1 and C chooses column 1 in Figure 5.50. The value of the game is 50, representing the percentage of the business that R gets.

In Example 1, we started with an applied situation and derived from it a matrix game. Actually, we can interpret any matrix as a game, as the following definition shows.

Definition Let *G* be an *m* X *n* matrix with entries g_{ij} for $i = 1, \ldots, m$ and $j = 1, \ldots, n$. Then *G* can be interpreted as the *pay-off matrix* of the following matrix game: player R (the *row* player) chooses any row *i*, and simultaneously player C (the *column* player) chooses any column *j;* the outcome of the game is that C pays to R an amount equal to g_{ij}. (If $g_{ij} < 0$, then this should be interpreted as R paying C an amount equal to $-g_{ij}$.)

Example 2: Consider the matrix in Figure 5.51 as a game. Thus, if R chooses row 1 and C chooses column 1, then C pays 5 units to R; if R chooses row 1 and C

		C Chooses 1	2
R Chooses	1	5	– 10
	2	2	0

Figure: 5.51

chooses column 2, then R pays 10 units to C; and so on. How should the players play this game?

Player R would like to get the 5 pay-off, and is tempted to play row 1. However, player C clearly prefers to play column 2, since each entry in it is lower than the corresponding entry in column 1. And since player R realises this, he will play row 2 to avoid the —10 pay-off. The optimal strategies then are "play row 2" for R, and "play column 2" for C. The value of the game is $g_{22} = 0$.

The solutions in the first two examples have the following in common. In each case, the value is an entry that is the minimum of its row and the maximum of its column. Such an entry is called a *saddle value.* When such a saddle value exists, it is always the value of the game, and the game is strictly determined. To see this, consider any game G with an entry $g_{ij} = v$ which is a saddle value. Then, since v is the minimum of row i, R can by playing row i assure that he will win at least u. And since v is the maximum of column j, C by playing column j can assure that R will not win more than v.

This justifies the definition:

Definition: Consider a matrix game with pay-off matrix G. Entry g_{ij} is said to be a *saddle value* of G if g_{ij} is simultaneously the *minimum* of the ith row and the *maximum* of the jth column. If matrix game G has a saddle value, it is said to be *strictly determined,* and *optimal strategies* for it are:

For player R: "Choose a row that contains a saddle value."

For player C: "Choose a column that contains a saddle value."

The *value* of the game is $v = g_{ij}$ where g_{ij} is any saddle-value entry. The game is *fair* if its value is zero.

In order to justify this definition, it must be shown that if there are two or more saddle values then they are all equal. A proof of this fact is outlined in Exercise 10. The next example illustrates it.

Example 3: Let us consider an extension of Example 1 in which the stores R and C are trying to locate in one of the three towns in Figure 5.52. We shall assume that if both stores locate in the same town they split all business equally, but if they locate in different towns then all the business in the town that

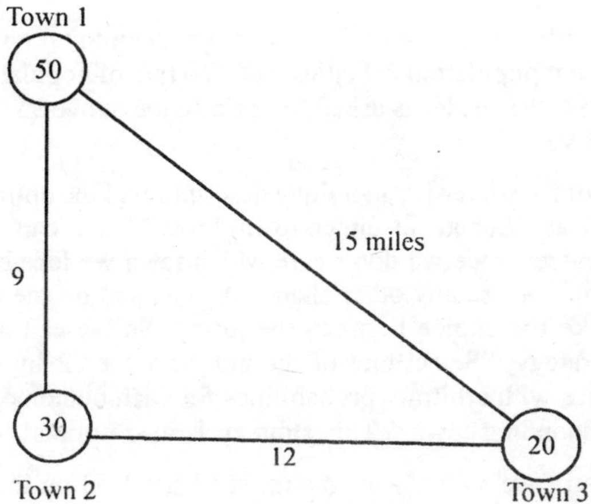

Figure: 5.52

doesn't have a store will go to the *closer* of the two stores. The percentages of people in each town are marked in the circles. The distances between the towns are marked on the lines connecting them.

The pay-off matrix for the resulting game is shown in Figure 5.53. In Exercise 13, the reader is asked to check that these entries are correct.

Each of the four 50 entries in the 2×2 matrix in the upper left-hand corner of Figure 5.53 is a saddle value of the matrix, since each is simultaneously the minimum of its row and maximum of its column. Note that

		store C locates in		
		1	2	3
Store R locates in	1	50	50	80
	2	50	50	80
	3	20	20	50

Figure: 5.53

the 50 entry in the lower right-hand corner is *not* a saddle value. Hence, the game is strictly determined, and optimal strategies are:

For store R: "Locate in either town 1 or town 2."

For store C: "Locate in either town 1 or town 2."

In a real-life location problem, one might want to take into account not only present populations of cities, but also rate of population growth. In Exercise 14, the reader is asked to criticise the above strategies from this point of view.

Instead of the somewhat indefinite description of the optimal strategy for player R as "Locate in either town 1 or 2," we can employ the following device: since we don't care which town we locate in, we can just flip a coin, or use any other chance device, and on the basis of the outcome make the choice between the towns. So we can also use the following strategy: "Select one of the numbers 1 or 2 by means of a random device with arbitrary probabilities for each outcome, and locate in the corresponding town." This strategy is also optimal.

Note that if we multiply the matrix in Figure 5.53 on the left by the vector (1,0,0), we get the first row; hence we shall use this vector to represent the strategy "Locate in town 1" for store R. Similarly, the strategy "Locate in town 2" is represented by the vector (0, 1,0), since multiplying the matrix on the left by it gives the second row. Then, the vector

$$(a,\ 1 - a,\ 0) = a(1,\ 0,\ 0) + (1 - a)(0,\ 1,0) \text{ for } 0 \leq a \leq 1$$

represents the strategy "Choose row 1 with probability a and row 2 with probability 1 — a."

Similarly, for store C, the column vectors

$$\begin{pmatrix}1\\0\\0\end{pmatrix},\ \begin{pmatrix}0\\1\\0\end{pmatrix},\ \text{and} \begin{pmatrix}a\\1-a\\0\end{pmatrix} \text{ for } 0 \leq a \leq 1$$

represents the strategies "Locate in town 1," "Locate in town 2," and "Locate in town 1 with probability a and in town 2 with probability 1 — a," respectively.

Example 4: Consider the game G whose matrix is in Figure 5.54. It is not hard to see that the game is strictly determined with value 1, and there are four saddle

Player c

	1	5	1	7
Player R	– 2	8	0	– 9
	1	12	1	3

Figure: 5.54

values. Optimal strategies are (1,0,0) and (0,0,1) for player R, and

$$\begin{pmatrix}1\\0\\0\\0\end{pmatrix} \text{and} \begin{pmatrix}0\\0\\1\\0\end{pmatrix}$$

for player C. The four ways we can pair optimal strategies for player R with those for player C give the four-saddle values. Besides the optimal strategies above, we have their convex combinations

$$a(1, 0, 0) + (1- a)(0, 0,1) = (a, 0, 1 - a),$$

which is optimal for R for any a satisfying $0 \leq a \leq 1$, and

$$a\begin{pmatrix}1\\0\\0\\0\end{pmatrix} + (1-a)\begin{pmatrix}0\\1\\0\\0\end{pmatrix} = \begin{pmatrix}a\\0\\1-a\\0\end{pmatrix},$$

which is optimal for player C for any a in the same range.

As the reader may have already found out for himself, not all matrix games are strictly determined. For instance, the two games shown in Figure 5.55 are not strictly determined. The solution of such games will be discussed in succeeding sections.

(a)

0	1
2	0

(b)

5	–2	3
–5	0	7
3	4	–1

Figure: 5.55

Exercises

1. Determine which of the games given below are strictly determined and which are fair. When the game is strictly determined, find optimal strategies for each player.

(a)

0	2
–1	4

(b)

5	0
0	2

(c)

3	1
4	0

(d)

1	−1
−1	1

(e)

3	1
−4	0

(f)

0	4
0	2

(g)

7	0
0	0

(h)

0	0
0	−7

(i)

0	0
0	0

(j)

1	1
1	1

[*Partial Ans.* (a) Strictly determined and fair; R play row 1, C play column 1; (b) non-strictly determined; (e) strictly determined but not fair; R play row 1, C play column 2; (j) strictly determined but not fair; both players can use any strategy.] 2. Find the value and all optimal strategies for the following games:

(a)

15	2	−3
6	5	7
−7	4	0

(b)

5	2	−1	−1
1	1	0	1
3	0	−3	7

(c)

0	5	6	−3
1	−1	2	3
1	2	3	4
−1	0	7	5

(d)

1	−12	6
0	−4	1
3	−7	2
3	−4	2
−5	−4	7

[*Ans.* (a) $v = 5$; (0, 1, 0); $\begin{pmatrix}0\\1\\0\end{pmatrix}$ (d) (0, a, 0, $1 - a$, 0), $\begin{pmatrix}0\\1\\0\end{pmatrix}$, $v = -4$].

3. Find the values of and all optimal strategies for the following games:

(a)

5	10	6	5
5	7	8	5
0	5	6	5

(b)

−2	0	−1
−5	7	8

(c)

0	0	1	0
1	0	0	0
1	0	1	0

(d)

3	2	3
6	2	7
5	1	4

$$[\textit{Ans.}\ (a)\ v = 5;\ (a,\ 1 - a,\ 0);\begin{pmatrix} a \\ 0 \\ 0 \\ 1-a \end{pmatrix};\ (d)\ v = 2;\ (a,\ 1 - a,\ 0);\begin{pmatrix} 0 \\ 1 \\ 0 \end{pmatrix}].$$

4. Each of two players shows one or two fingers (simultaneously) and C pays to R a sum equal to the total number of fingers shown. Write the game matrix. Show that the game is strictly determined, and find the value and optimal strategies.

5. Each of two players shows one or two fingers (simultaneously) and C pays to R an amount equal to the total number of fingers shown, while R pays to C an amount equal to the product of the numbers of fingers shown. Construct the game matrix (the entries will be the net gain of R), and find the value and the optimal strategies.

   [*Ans.* $v = 1$, R must show one finger, C may show one or two.]

6. Show that a strictly determined game is fair if and only if there is a zero entry such that all entries in its row are non-negative and all entries in its column are non-positive.

7. Consider the game

$G =$

2	5
–1	a

   (a) Show that G is strictly determined regardless of the value of a.

   (b) Find the value of G. [*Ans.* 2.]

   (c) Find optimal strategies for each player.

   (d) If a = 1,000,000, obviously R would like to get it as his pay-off.

   Is there any way he can assure himself of obtaining it? What would happen to him if he tried to obtain it?

   (c) Show that the value of the game is the most that R can assure for himself.

8. Consider the matrix game

$G =$

a	a
c	d

Show that G is strictly determined for every set of values for a, c, and d. Show that the same result is true if two entries in a given column are equal.

9. Find necessary and sufficient conditions that the game

$G =$

a	0
0	b

should be strictly determined. [*Hint:* These will be expressed in terms of relations among the numbers a and b and the number zero.]

10. (a) Show that if there are two saddle values in the same row, then they are equal.

    (b) Show that if there are two saddle values in the same column, then they are equal.

    (c) If g_{ij} and g_{hk} are saddle values in different rows and columns, show that $g_{ij} = g_{ik}$. Also show $g_{ik} = g_{hk}$.

    (d) Prove that $g_{ij} = g_{hk}$.

11. Two companies, one large and one small, manufacturing the same product, wish to build a new store in one of four towns located on a given highway. If we regard the total population of the four towns as 100 per cent, the distribution of population and distances between towns are as shown:

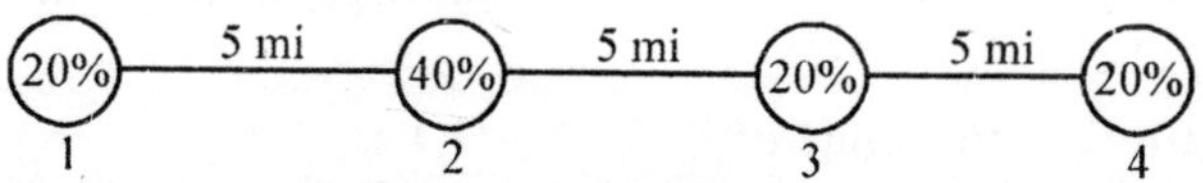

Assume that if the large company's store is nearer a town, it will capture 80 per cent of the business; if both stores are equally distant, then the large company will capture 60 per cent of the business; and if the small store is nearer, then the large company will capture 40 per cent of the business.

(a) Set up the matrix of the game.

(b) Test for dominated rows and columns, that is, rows or columns that will never be used by a player who plays optimally.

(c) Find optimal strategies and the value of the game and interpret your results.

12. Rework Exercise 11 if the percentages of business captured by the large company are 90, 75, and 60, respectively.

[*Ans.* Both companies should locate in town 2; the large company captures 60 per cent of the business.]

13. Show that the entries in Figure 5.53 are correct.
14. In the store location of Example 3 how do the optimal strategies change if the population of town 1 becomes 51 per cent and the population of town 2 becomes 29 per cent of the total? How might they change if town 2 is growing much faster than town 1?
15. Show that the following game is always strictly determined for non-negative a and any values of the parameters b, c, d, and e.

$2a$	a	$3a$
b	$-a$	c
d	$-2a$	e

16. For what values of a is the following game strictly determined?

[*Ans.* $-1 \leq a \leq 2$.]

a	6	2
-1	a	-7
-2	4	a

Matrix Games

As we saw in the numerical examples of the previous section, some matrix games are non-strictly determined; that is, they have no entry that is simultaneously a row minimum and a column maximum. We can characterise non-strictly determined 2×2 matrix games as follows:

Theorem: The matrix game

G =

a	b
c	d

is non-strictly determined if and only if one of the following two conditions is satisfied:

(i) $a < b$, $a < c$, $d < b$, and $d < c$.

(ii) $a > b$, $a > c$, $d > b$, and $d > c$.

(These equations mean that the two entries on one diagonal of the matrix must each be greater than each of the two entries on the other diagonal.)

Proof: If either of the conditions (i) or (ii) holds, it is easy to check that no entry of the matrix is simultaneously the minimum of the row and the maximum of the column in which it occurs; hence the game is not strictly determined.

To prove the other half of the theorem, recall that, by Exercise 8 of the last section, if two of the entries in the same row or the same column of G are equal, the game is strictly determined; hence we can assume that no two entries in the same row or the same column are equal. Suppose now that $a < b$; then $a < c$ or else a is a row minimum and a column maximum; then also $c > d$ or else c is a row minimum and a column maximum; then also $d < b$ or else d is a row minimum and a column maximum. Hence, the assumption $a < b$ leads to case (i) above.

In a similar manner the assumption $a > b$ leads to case (ii). This completes the proof of the theorem.

Example 1: Jones and Smith play the following game: Jones conceals either a \$1 or a \$2 bill in his hand; Smith guesses 1 or 2, winning the bill if he guesses the number. If we make Jones player R (the row player) and Smith player C, the matrix of the game is as in Figure 5.56. Because the game satisfies condition (i) in the theorem above, the game is non-strictly determined. Later we shall solve it.

		Player C Smith guesses 1	2
Player R Jones chooses	\$1 bill	–1	0
	\$2 bill	0	–2

Figure: 5.56

Example 2: Mr. Sub works for Mr. Super and frequently must advise him on the acceptability of certain projects. Whenever Mr. Sub can make a clear judgement about a given project, he does so honestly. But when he has no reason to either accept or reject a given project, he tries to agree with Mr. Super. If he manages to agree with him he gives himself 10 points; if he is unfavourable when his boss is favourable, he credits himself with 0 points; but when he is favourable and his boss is unfavourable (the worst case), he loses 50 points. The matrix of the game is given in Figure 5.57. Since the matrix in Figure 5.57 satisfies condition (ii) of the theorem, it is not strictly determined.

How should one play a non-strictly determined game? We must first convince ourselves that no single choice is clearly optimal for either player.

		Player C Mr. Super's opinion Favourable	 Unfavourable
Player R Mr. Sub's opinion	Favourable	10	–50
	Unfavourable	0	10

Figure: 5.57

In Example 1, R would like to get one of the 0 pay-offs. But if he always chooses \$1 and C finds this out, C can win \$1 by guessing 1. And if R always chooses \$2, then C can win \$2 by guessing 2. Similarly, if C always guesses 1 or always guesses 2, and R finds this out, then R can always get 0. So our first result is that each player must, in some way, prevent the other player from finding out which choice of alternatives he is going to make.

We also note that for a single play of a non-strictly determined 2 × 2 game there is no difference between the two strategies, as long as one's strategy is not guessed by the opponent. Let us now consider several plays of the game. What should R do? Clearly, he should not choose the same row all the time, or C will be able to notice and profit by it. Rather, R should choose sometimes one row, sometimes the other. Our key question then is "How often should R choose each of his alternatives?" In Example 1, it seems reasonable that player R (Jones) should choose the \$1 bill about twice as often as the \$2 bill, because his losses, if Smith guesses correctly, are half as much. In what order should he do this? For instance, should he select the \$1 bill twice in a row and then the \$2 bill? That is dangerous, because if player C (Smith) notices the pattern, he can gain by knowing just what R will do next. Thus, we see that R should choose the \$1 bill two-thirds of the time, but according to some unguessable pattern. The only safe way of doing this is to play it two-thirds of the time at random. He could, for instance, roll a die (without letting C see it) and choose the \$1 if 1 through 4 turns up, the \$2 if 5 or 6 turns up. Then his opponent cannot guess what the actual decision will be, since

R himself won't know it. We conclude that a rational way of playing is for each player to *mix* his strategies, selecting sometimes one, sometimes the other; and these strategies should be selected at random, according to certain fixed ratios (probabilities) of selecting each.

By a *mixed strategy* in a 2 × 2 game for player R we shall mean a command of the form "Play row 1 with probability p_1 and play row 2 with probability p_2," where we assume that $p_1 \geq 0$ and $p_2 \geq 0$ and $p_1 + p_2 = 1$. Similarly, a mixed strategy for player C is a command of the form "Play column 1 with probability q_1 and play column 2 with probability q_2" where $q_1 \geq 0, q_2 \geq 0$, and $q_1 + q_2 = 1$. A mixed-strategy vector for player R is the probability row vector (p_1, p_2), and a mixed-strategy vector for player C is the probability column vector $\begin{pmatrix} q_1 \\ q_2 \end{pmatrix}$.

Examples of mixed strategies are $\left(\frac{1}{2}, \frac{1}{2}\right)$ and $\begin{pmatrix} \frac{1}{5} \\ \frac{4}{5} \end{pmatrix}$. The reader may wonder how a player could actually play one of these strategies. The mixed strategy $\left(\frac{1}{2}, \frac{1}{2}\right)$ is easy to realise, since it can be realised by flipping a coin and choosing one alternative if heads turns up and the other alternative if tails turns up. The mixed strategy $\begin{pmatrix} \frac{1}{5} \\ \frac{4}{5} \end{pmatrix}$ is more difficult to realise, since no chance device in common use gives these probabilities.

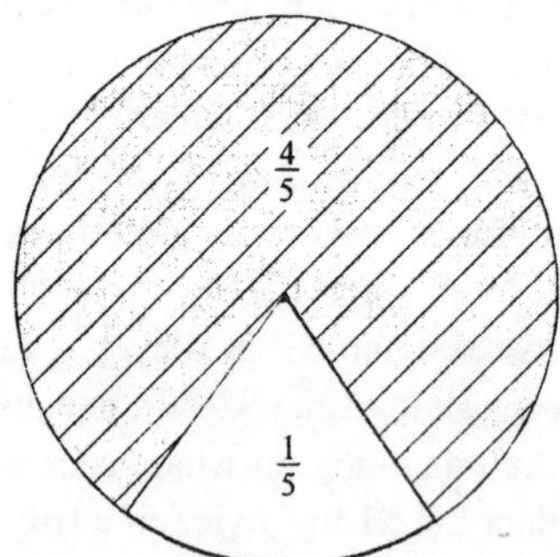

Figure: 5.58

However, suppose a pointer is constructed with a card that is $\frac{4}{5}$ shaded and $\frac{1}{5}$ unshaded, as in Figure 5.58, and C simply spins the pointer (without letting R see it, of course!). Then, if the pointer stops on the unshaded part he plays the first column, and if it stops on the shaded part he plays the second column, thus realising the desired strategy. By varying the proportion of shaded area on the card, other mixed strategies can conveniently be realised. An equally effective and less mechanical device for realising a given mixed strategy is to use a table of random digits. For the strategy $\begin{pmatrix} \frac{1}{5} \\ \frac{4}{5} \end{pmatrix}$, for example, we could let the digits 0 and 1 represent a play of column 1, and the remaining digits a play of column 2.

We now want to define what we shall mean by a solution to an $m \times n$ matrix game.

Definition: Let G be an $m \times n$ matrix with entries g_{ij}. An m-component row vector p is a *mixed-strategy vector* for player R if it is a probability vector; similarly, an n-component column vector q is a *mixed-strategy vector for* C if it is a probability vector. (Recall that a probability vector is one with non-negative entries whose sum is 1.) Let v be a number, let e be an m-component row vector all of whose entries are 1, and let f be an n-component column vector all of whose entries are 1. It follows that the vectors ve and vf are

$$ve = \underbrace{(v, v, \ldots, v)}_{m \text{ components}} \quad \text{and} \quad vf = \left.\begin{pmatrix} v \\ v \\ \vdots \\ v \end{pmatrix}\right\} n \text{ components}$$

Then v is the *value* of the matrix game G and p^0 and q^0 are *optimal strategies* for the players if and only if the following inequalities hold:

$$p^0 G \geq ve, \tag{1}$$

$$Gq^0 \geq vf. \tag{2}$$

Example 3: In Example 1 of the previous section we had the matrix:

$$G = \begin{pmatrix} 50 & 60 \\ 40 & 50 \end{pmatrix}.$$

We found that the value of this game was $v = 50$ and that optimal strategies were for R to choose row 1, which corresponds to the mixed-strategy vector $p^0 = (1, 0)$, and for C to choose column 1, which corresponds to the mixed-strategy vector $q^0 = \begin{pmatrix} 1 \\ 0 \end{pmatrix}$. Carrying out the calculations in (1) and (2), we have

$$p^0G = (1,0)\begin{pmatrix} 50 & 60 \\ 40 & 50 \end{pmatrix} = (50,60) \geq (50,50) = 50(1,1) = ve$$

and

$$Gq^0 = \begin{pmatrix} 50 & 60 \\ 40 & 50 \end{pmatrix}\begin{pmatrix} 1 \\ 0 \end{pmatrix} = \begin{pmatrix} 50 \\ 40 \end{pmatrix} \leq \begin{pmatrix} 50 \\ 50 \end{pmatrix} = 50\begin{pmatrix} 1 \\ 1 \end{pmatrix} = vf.$$

Let us return now to the non-strictly determined 2×2 game. Consider the non-strictly determined game

$$G = \begin{array}{|c|c|} \hline a & b \\ \hline c & d \\ \hline \end{array}$$

Having argued, as above, that the players should use mixed strategies in playing a non-strictly determined game, it is still necessary to decide how to choose an optimal mixed strategy.

If R chooses a mixed strategy $p = (p_1, p_2)$ and (independently) C chooses a mixed strategy $q = \begin{pmatrix} q_1 \\ q_2 \end{pmatrix}$, then player R obtains the pay-off a with probability p_1q_1; he obtains the pay-off b with probability p_1q_2; he obtains c with probability p_2q_2; and he obtains d with probability p_2q_2; hence his mathematical expectation is given by the expression

$$ap_1q_1 + bp_1q_2 + cp_2q_1 + dp_2q_2.$$

By a similar computation, one can show that player C's expectation is the negative of this expression.

To justify this definition we must show that if v, p^0, q^0 exist for G, each player can guarantee himself an expectation of v. Let q be any strategy for C. Multiplying (1) on the right by q, we get

$$p^0Gq \geq (v, v)q = v,$$

which shows that, regardless of how C plays, R can assure himself of an expectation of at least v. Similarly, let p be any strategy vector for R. Multiplying (2) on the left by p, we obtain

$$pGq^0 \leq p\begin{pmatrix} v \\ v \end{pmatrix} = v,$$

which shows that, regardless of how R plays, C can assure himself of an expectation of at most v. It is in this sense that p^0 and q^0 are optimal.

It follows further that, if both players play optimally, then R's expectation is exactly v and C's expectation is exactly v. Hence, we call v the (expected) value of the game.

We must now see whether there are strategies p^0 and q^0 for the game G. For a 2 × 2 non-strictly determined game the following formulas provide the solution:

$$p_1^0 = \frac{d-c}{a+d-b-c}, \tag{3}$$

$$p_2^0 = \frac{a-b}{a+d-b-c}, \tag{4}$$

$$q_1^0 = \frac{d-b}{a+d-b-c}, \tag{5}$$

$$q_2^0 = \frac{a-c}{a+d-b-c}, \tag{6}$$

$$v = \frac{ad-bc}{a+d-b-c}. \tag{7}$$

It is an easy matter to verify that formulas (3)-(7) satisfy conditions (l)-(2). Actually, the inequalities in (1) and (2) become equalities in this simple case, a fact that is not true in general for non-strictly determined games of larger size.

The denominator in each formula is the difference between the sums of the entries on the two diagonals. Since, for a non-strictly determined game, the entries on one diagonal must be larger than those on the other, the denominator cannot be zero.

Let us use these formulas to solve the examples mentioned earlier.

Applying formulas (3)-(7) to the matrix in Figure 5.56, we have

$$p_1^0 = \frac{-2-0}{-1-2-0-0} = \frac{2}{3},\ p_2^0 = \frac{-1-0}{-3} = \frac{1}{3},$$

$$q_1^0 = \frac{-2-0}{-3} = \frac{2}{3},\ q_2^0 = \frac{-1-0}{-3} = \frac{1}{3},\ v = \frac{(-1)(-2)-0}{-3} = -\frac{2}{3}.$$

Thus, the game is biased in favour of player C, since $v = -\frac{2}{3}$, and optimal strategies are

$$p^0 = \left(\tfrac{2}{3}, \tfrac{1}{3}\right) \text{ and } q^0 = \begin{pmatrix} \frac{2}{3} \\ \frac{1}{3} \end{pmatrix}.$$

Both Jones and Smith should select their first alternative two-thirds of the time, according to some random pattern.

Example 2: Let us apply the formulas (3)-(7) to the matrix in Figure 5.57. We obtain

$$a + d - c - b = 10 + 10 - 0 + 50 = 70,$$

so that:

$$p_1^0 = \frac{10-0}{70} = \frac{1}{7}, \quad p_2^0 = \frac{10+50}{70} = \frac{6}{7},$$

$$q_1^0 = \frac{10+50}{70} = \frac{6}{7}, \quad q_2^0 = \frac{10-0}{70} = \frac{1}{7}, \quad v = \frac{10 \cdot 10 - 0}{70} = \frac{10}{7}.$$

Notice that the game is biased in favour of Mr. Sub, not his boss Mr. Super! Also Mr. Sub's optimal strategy is to have an *unfavourable* opinion 6 out of 7 times, while Mr. Super's optimal strategy is to have a *favourable* opinion 6 out of 7 times! Thus, if this game is at all realistic, a subordinate should be much more critical than his superior when judging situations in which there is no clear-cut reason to either accept or reject a project. The conclusion is based on game-theory analysis, not on the two persons' relative ages, experience, and so on.

We conclude this section by proving three theorems that characterise the value and optimal strategies of a game.

Theorem: If G is a matrix game that has a value and optimal strategies, then the value of the game is unique.

***Proof*:** Suppose that v and w are two different values for the game G. Then let p^0 and q^0 be optimal mixed-strategy vectors associated with the value v such that

$$p^0 G \geq ve, \tag{a}$$

$$Gq^0 \leq vf. \tag{b}$$

Similarly, let p^1 and q^1 be optimal mixed-strategy vectors associated with the value w such that

$$p^1 G \geq we, \tag{c}$$

$$Gq^1 \leq wf. \tag{d}$$

If we now multiply (a) on the right by q^1, we get $p^0Gq^1 \geq (ve)q^1 = v$. In the same way, multiplying (d) on the left by p^0 gives $p^0Gq^1 \leq w$. The two inequalities just obtained show that $w \geq v$.

Next we multiply (b) on the left by p^1 and (c) on the right by q^0, obtaining $v \geq p^1Gq^0$ and $p^1Gq^0 \geq w$, which together imply that $v \geq w$.

Finally we see that $v \leq w$ and $v \geq w$ imply together that $v = w$—that is, the value of the game is unique.

Theorem: If G is a matrix game with value v and optimal strategies p^0 and q^0, then $v = p^0Gq^0$.

Proof: By definition v, p^0, and q^0 satisfy

$p^0G \geq ve$ and $Gq^0 \leq vf$.

Multiplying the first of these inequalities on the right by q^0, we get $p^0Gq^0 \geq v$. Similarly, multiplying the second inequality on the left by p^0, we obtain $p^0Gq^0 \leq v$. These two inequalities together imply that $v = p^0Gq^0$, concluding the proof.

The theorems just proved are important because they permit us to interpret the *value* of a game as an *expected value*. Briefly the interpretation is the following: If the game G is played repeatedly and if each time it is played player R uses the mixed strategy p^0 and player C uses the mixed strategy q^0, then the value v of G is the expected value of the game for R. The law of large numbers implies that, if the number of plays of G is sufficiently large, then the average value of R's winnings will (with high probability) be arbitrarily close to the value v of the game G.

As an example, let G be the matrix of the game of matching pennies:

$G =$

1	–1
–1	1

Using the formulas above, we find that optimal strategies in this game are for R to choose each row with probability $\frac{1}{2}$ and for C to choose each column with probability $\frac{1}{2}$. The value of G is zero. Notice that the only two pay-offs that result from a single play of the game are +1 and –1, neither of which is equal to the value of the game. However, if the game is played repeatedly, the average value of R's pay-offs will approach zero, which is the value of the game.

Theorem: If G is a game with value v and optimal strategies p^0 and q^0, then v is the largest expectation that R can assure for himself. Similarly, v is the smallest expectation that C can assure for himself.

Proof: Let p be any mixed-strategy vector of R and let q^0 be an optimal strategy for C; then multiply the equation $Gq^0 \leq vf$ on the left by p, obtaining $pGq^0 \leq v$. The latter equation shows that, if C plays optimally, the most that R can assure for himself is v. Now let p^0 be optimal for R; then, for every q, $p^0Gq \geq v$, so that R can actually assure himself of an expectation of v. The proof of the other statement of the theorem is similar.

The theorem above gives an intuitive justification to the definition of value and optimal strategies for a game. Thus, the value is the "best" that a player can assure himself, and optimal strategies are the means of assuring this "best."

Exercises

1. Find the optimal strategies for each player and the values of the following games:

(a)

1	2
3	4

(b)

1	0
–1	2

(c)

2	3
1	4

(d)

15	3
–1	2

(e)

7	–6
5	8

(f)

3	15
–1	10

[*Ans.* (a) $v = 3$; $(0, 1)$; $\begin{pmatrix}1\\0\end{pmatrix}$; (b) $v = \frac{1}{2}$; $\left(\frac{3}{4}, \frac{1}{4}\right)$; $\begin{pmatrix}\frac{1}{2}\\\frac{1}{2}\end{pmatrix}$;

(d) $v = 3$; $(1,0)$; $\begin{pmatrix}1\\0\end{pmatrix}$; (e) $v = \frac{43}{8}$; $\left(\frac{3}{16}, \frac{13}{16}\right)$; $\begin{pmatrix}\frac{7}{8}\\\frac{1}{8}\end{pmatrix}$.]

2. Set up the ordinary game of matching pennies as a matrix game. Find its value and optimal strategies. How are the optimal strategies realised in practice by players of this game?

3. A version of two-finger Morra is played as follows: Each player holds up either one or two fingers; if the sum of the number of fingers shown is even, player R gets the sum, and if the sum is odd, player C gets it.

   (a) Show that the game matrix is

		Player C 1	Player C 2
Player R	1	2	–3
	2	–3	4

   (b) Find optimal strategies for each player and the value of the game.

[*Ans.* $\left(\frac{7}{12}, \frac{5}{12}\right), \begin{pmatrix} \frac{7}{12} \\ \frac{5}{12} \end{pmatrix}, v = -\frac{1}{12}.$]

4. Rework Exercise 3 if player C gets the even sum and player R gets the odd sum.

5. Let G be the matrix in Figure 5.51. With $v = 0$, $p^0 = (0, 1)$, and $q^0 = \begin{pmatrix} 0 \\ 1 \end{pmatrix}$, show that formulas (1) and (2) are satisfied.

6. If

$$G = \begin{array}{|c|c|} \hline a & b \\ \hline c & d \\ \hline \end{array}$$

   is non-strictly determined, prove that it is fair if and only if $ad = bc$.

7. In formulas (3)-(7) prove that $p_1 > 0$, $p_2 > 0$, $q_1 > 0$, and $q_2 > 0$. Must v be greater than zero?

8. Find necessary and sufficient conditions that the game

$$G = \begin{array}{|c|c|} \hline a & 0 \\ \hline 0 & b \\ \hline \end{array}$$

   be non-strictly determined. Find optimal strategies for each player and the value of G if it is non-strictly determined.

[*Ans.* a and b must be both positive or both negative. $p_1 = b/(a + b)$; $p_2 = a/(a + b)$; $q_1 = b/(a + b)$; $q_2 = a/(a + b)$; $v = ab/(a + b)$.]

9. Suppose that player R tries to find C in one of three towns X, Y, and Z. The distance between X and Y is five miles, the distance between Y and Z is five miles, and the distance between Z and X is ten miles. Assume that R and C can each go to one and only one of the three towns and that if they both go to the same town R "catches" C; otherwise C "escapes." Credit R with ten points if he catches C, and credit C with a number of points equal to the number of miles he is away from R if he escapes.
   (a) Set up the game matrix.
   (b) Show that both players have the same optimal strategy, namely, to go to towns X and Z with equal probabilities and to go to town Y with probability $\frac{1}{4}$.
   (c) Find the value of the game.
10. Verify that formulas (3)-(7) satisfy conditions (1) and (2).
11. Consider the (symmetric) game whose matrix is

$$G = \begin{bmatrix} 0 & -a & -b \\ a & 0 & -c \\ b & c & 0 \end{bmatrix}$$

   (a) If a and b are both positive or both negative, show that G is strictly determined.
   (b) If b and c are both positive or both negative, show that G is strictly determined.
   (c) If $a > 0$, $b < 0$, and $c > 0$, show that an optimal strategy for player R is given by

$$\left(\frac{c}{a-b+c}, \frac{-b}{a-b+c}, \frac{a}{a-b+c}\right).$$

   (d) In part (c) find an optimal strategy for player C.
   (e) If $a < 0$, $b > 0$, and $c < 0$, show that the strategy given in (c) is optimal for R. What is an optimal strategy for player C?
   (f) Prove that the value of the game is always zero.

12. In a well-known children's game each player says "stone" or "scissors" or "paper." If one says "stone" and the other "scissors," then the former wins a penny. Similarly, "scissors" beats "paper," and "paper" beats "stone." If the two players name the same item, then the game is a tie.

    (a) Set up the game matrix.

    (b) Use the results of Exercise 10 to solve the game.

13. In Exercise 13 let us suppose that the payments are different in different cases. Suppose that when "stone breaks scissors" the payment is one cent; when "scissors cut paper" the payment is two cents; and when "paper covers stone" the payment is three cents.

    (a) Set up the game matrix.

    (b) Use the results of Exercise 9 to solve the game.

    [*Ans.* $\frac{1}{3}$ "stone," $\frac{1}{2}$ "scissors," $\frac{1}{6}$ "paper"; $v = 0$.]

14. A strictly determined $m \times n$ matrix game G contains a saddle entry g_{ij} that is simultaneously the minimum of row i and the maximum of column j.

    (a) Show that by rearranging rows and columns (if necessary) we can assume that g_{11} is a saddle value.

    (b) Let $v = g_{11}$ and p^0 and q^0 be probability vectors with first component equal to 1 and all other components equal to 0. Show that these quantities satisfy (1) and (2).

15. Verify that the strategies $p^0 = \left(\frac{1}{3}, \frac{1}{3}, \frac{1}{3}\right)$ and

$$q^0 = \begin{pmatrix} \frac{1}{3} \\ \frac{1}{3} \\ \frac{1}{3} \end{pmatrix}$$

are optimal in the game G whose matrix is

$$G = \begin{array}{|c|c|c|} \hline 1 & 0 & 0 \\ \hline 0 & 1 & 0 \\ \hline 0 & 0 & 1 \\ \hline \end{array}$$

What is the value of the game?

16. Generalise the result of Exercise 12 to the game G whose matrix is the $n \times n$ identity matrix.

17. Consider the following game:

$G =$

a	0	0
0	b	0
0	0	c

(a) If a, b, and c are not all of the same sign, show that the game is strictly determined with value zero.

(b) If a, b, and c are all of the same sign, show that the vector

$$\left(\frac{bc}{ab+bc+ca}, \frac{ca}{ab+bc+ca}, \frac{ab}{ab+bc+ca}\right)$$

is an optimal strategy for player R.

(c) Find player C's optimal strategy for case b.

(d) Find the value of the game for case b, and show that it is positive if a, b, and c are all positive, and negative if they are all negative.

18. Suppose that the entries of a matrix game are rewritten in new units (e.g., dollars instead of cents). Show that the monetary value of the game has not changed.

19. Consider the game of matching pennies whose matrix is

1	−1
−1	1

If the entries of the matrix represent gains or losses of one penny, would you be willing to play the game at least once? If the entries represent gains or losses of one dollar, would you be willing to play the game at least once? If they represent gains or losses of one million dollars, would you play the game at least once? In each of these cases show that the value is zero and optimal strategies are the same. Discuss the practical application of the theory of games in the light of this example.

Simplex Method for Solving Matrix Games

We have so far restricted our attention to examples of Matrix Games that were simple enough to be solved by unformalised computations.

However, games of realistic size frequently lead to very large matrices for which these simple techniques are not adequate. A clue to a general technique may be found in the fact that the row player is a maximising player while the column player is a minimising player.

There are several ways of showing that a matrix game is equivalent to a linear programme. We choose a very simple approach here, based on the fact, that every matrix game is equivalent to one in which all entries are positive and hence whose value is positive.

Besides finding an equivalent linear programming problem, we shall give a proof, based on the duality theorem of linear programming, that every matrix game has a solution. And we shall present a simplex format suitable for the solution of any matrix game.

Let G be an $m \times n$ matrix game and let E be the $m \times n$ matrix all of whose entries are l's. The second theorem of previous section states that G and $G + kE$ have the same optimal strategies, and the value of the second game is k plus the value of the first game. Hence, if we start with any game G we can replace it by a game G' all of whose entries are positive, and which has the same optimal strategies. For instance, to get G' we could add 1 minus the most negative entry in G to every entry in G

Thus, without loss of generality we let G be a positive matrix game. We also let e be the n-component row vector all of whose entries are l's, and let f be the m-component column vector all of whose entries are l's. Let z be an m-component row vector and x an n-component column vector.

We now consider the following dual linear programming problems:

$$(1) \quad \begin{cases} Minimise\ zf \\ subject\ to: \\ \quad zG \geq e \\ \quad z \geq 0 \end{cases} \qquad \begin{cases} Minimise\ ex \\ subject\ to: \\ \quad Gx \leq f \\ \quad x \geq 0. \end{cases}$$

Note that $x = 0$ satisfies the constraints of the maximising problem; also, because the entries of G and f are positive, there is at least one non-zero x vector that will satisfy these constraints. Moreover, the set of feasible x vectors is bounded. Because of these facts, and because e has all positive entries, the maximising problem has solution x^0 such that

$ex^0 > 0$. Hence, by the duality theorem, the minimum problem has a solution x^0 and

$$t = z^0Gx^0 = z^0f = ex^0 > 0.$$

We now set

$$p^0 = \frac{z^0}{t}, \quad q^0 = \frac{x^0}{t}, \text{ and } v = \frac{1}{t},$$

and observe that p^0 and q^0 are probability vectors.

It is easy to see that p^0 is an optimal strategy for player R in G, since x^0 satisfies the constraints of the minimising problem, and hence

$$p^0G = \frac{z^0G}{t} \geq \frac{e}{t} = ve.$$

In Exercise 1 the reader is asked to show similarly that $Gq^0 \leq vf$. We summarise these results in the following theorem:

Theorem: (a) Solving the matrix game G with positive entries can be accomplished by solving the dual linear programming problems (1).

(b) Every matrix game has at least one solution; solutions to such games can be found by the simplex method.

Actually, it is not necessary that the matrix game be positive in order that the simplex method work. It is enough that its value be positive. However, in Exercise 3 the reader is asked to work a specific example for which the linear programming problem as described above has no solution if applied to a game with zero value.

Before proceeding to specific examples, let us outline the procedure to be followed in setting up a matrix game for solution by the simplex method.

1. Set up the matrix of the game.
2. Check to see whether the game is strictly determined; if so, the solution is already obtained.
3. Check for row and column dominance and remove dominated rows and columns.

4. Make certain that the value of the game is positive. It is sufficient for this to add 1 minus the most negative matrix entry to every entry of the matrix. Let k be the amount added, if any, to each matrix entry.
5. Let G be the matrix of the resulting game; suppose it is m × n. Construct the following matrix tableau:

G	f
e	0

6. Carry out the steps of the simplex algorithm until all indicators are non-positive. Determine the solutions z^0 and x^0 to the dual linear programming problems, and let $t = z^0 f = ex^0$. We know $t > 0$.
7. The solutions to the original matrix game are given by

$$p^0 = \frac{z^0}{t}, \quad q^0 = \frac{x^0}{t}, \text{ and } v = \frac{1}{t} - k.$$

(If dominated rows or columns were removed from the game, the strategy vectors may have to be extended by the addition of some zero components).

Example 1: We know that the matching-pennies game is fair—that is, it has value zero. To make its value positive, we add $k = 2$ to each matrix entry, yielding the following game:

3	1
1	3

Obviously this game is not strictly determined and it does not have dominated rows or columns. We set up the simplex tableau and solve it as shown in Figure 5.62(a)-(c). (Note that we have called the variables on the left z_x and z_2 instead of v_1 and v_2, since we are now using v for the value of the game.) From the final tableau in Figure 5.62(c) we can see that the value of the game is 2 (the reciprocal of $t = \frac{1}{2}$), so that the value of the matching-pennies game is 0, which we know already. Also, optimal strategies are

$$p^0 = \frac{z^0}{t} = \left(\tfrac{1}{2}, \tfrac{1}{2}\right) \quad \textit{and} \quad q^0 = \frac{x^0}{t} = \begin{pmatrix} \frac{1}{2} \\ \frac{1}{2} \end{pmatrix},$$

which we had discovered earlier.

Example 2: Let us solve the two-finger Morra game. To convert the game into one with positive value let us add 3 to each entry of the matrix,

	x_1	x_2	-1	
z_1	$\textcircled{3}$	1	1	$= -y_1$
z_2	1	3	1	$= -y_2$
-1	1	1	0	$= f$
	$= u_1$	$= u_2$	$= g$	

(a)

	y_1	x_2	-1	
u_1	$\frac{1}{3}$	$\frac{1}{3}$	$\frac{1}{3}$	$= -x_1$
z_2	$-\frac{1}{3}$	$\textcircled{\frac{8}{3}}$	$\frac{2}{3}$	$= -y_2$
-1	$-\frac{1}{3}$	$\frac{2}{3}$	$\frac{1}{3}$	$= f$
	$= z_1$	$= u_2$	$= g$	

(b)

	y_1	y_2	-1	
u_1	$\frac{3}{8}$	$-\frac{1}{3}$	$\frac{1}{4}$	$= -x_1$
u_2	$-\frac{1}{3}$	$\frac{3}{8}$	$\frac{1}{4}$	$= -x_2$
-1	$-\frac{1}{4}$	$-\frac{1}{4}$	$\frac{1}{2}$	$= f$
	$= z_1$	$= z_2$	$= g$	

(c)

Figure 5.62

noting that this will give two zeros in the resulting game matrix. These zeros will simplify the simplex calculations. The game matrix now is

5	0
0	7

Figure 5.63(a)-(c) shows the initial and two subsequent simplex tableaus. The value of the game, from Figure 63(c), is $\frac{35}{12}$, which means that the value of the original game is $\frac{35}{12} - 3 = -\frac{1}{12}$.

Example 3: Consider the following game: R and C simultaneously display 1, 2, or 3 pennies. If both show the same number of pennies, no money is exchanged; but if they show different numbers of pennies, R gets odd sums and C gets

	x_1	x_2	-1	
z_1	(5)	0	1	$= -y_1$
z_2	0	7	1	$= -y_2$
-1	1	1	0	$= f$
	$= u_1$	$= u_2$	$= g$	

(a)

	y_1	x_2	-1	
u_1	$\frac{1}{5}$	0	$\frac{1}{5}$	$= -x_1$
z_2	0	(7)	1	$= -y_2$
-1	$-\frac{1}{5}$	1	$\frac{1}{5}$	$= f$
	$= z_1$	$= u_2$	$= g$	

(b)

	y_1	y_2	-1	
u_1	$\frac{1}{5}$	0	$\frac{1}{5}$	$= -x_1$
u_2	0	$\frac{1}{7}$	$\frac{1}{7}$	$= -x_2$
-1	$-\frac{1}{5}$	$-\frac{1}{7}$	$\frac{12}{35}$	$= f$
	$= z_1$	$= z_2$	$= g$	

(c)

Figure 5.63

even sums. The matrix of the game is

		C shows	
	1	2	3
1	0	3	–4
R shows 2	3	0	5
3	–4	5	0

Since the second row has all non-negative entries, the game is, if anything, in R's favour. And if R plays the first two rows with equal weight, his expectation is positive. Hence, the value of the game is positive, and we do not have to add anything to the matrix entries. The simplex calculations needed to solve the game are shown in Figure 5.64(a)-(d). From this we see that the value of the game is $1/7 = \frac{10}{7}$, and that optimal strategies are

$$p^0 = \frac{z^0}{t} = \left(\tfrac{5}{14}, \tfrac{4}{7}, \tfrac{1}{14}\right) \text{ and } q^0 = \frac{x^0}{t} = \begin{pmatrix} \frac{5}{14} \\ \frac{4}{7} \\ \frac{1}{14} \end{pmatrix}.$$

	x_1	x_2	x_3	-1	
z_1	0	3	-4	1	$= -y_1$
z_2	③	0	5	1	$= -y_2$
z_3	-4	5	0	1	$= -y_3$
-1	1	1	1	0	$= f$
	$= u_1$	$= u_2$	$= u_3$	$= g$	

(a)

	y_2	x_2	x_3	-1	
z_1	0	③	-4	1	$= -y_1$
u_1	$\frac{1}{3}$	0	$\frac{5}{3}$	$\frac{1}{3}$	$= -x_1$
z_3	$\frac{4}{3}$	5	$\frac{20}{3}$	$\frac{7}{3}$	$= -y_3$
-1	$-\frac{1}{3}$	1	$\frac{2}{3}$	$\frac{1}{3}$	$= f$
	$= z_2$	$= u_2$	$= u_3$	$= g$	

(b)

	y_2	y_1	x_3	-1	
u_2	0	$\frac{1}{3}$	$-\frac{4}{3}$	$\frac{1}{3}$	$= -x_2$
u_1	$\frac{1}{3}$	0	$\frac{5}{3}$	$\frac{1}{3}$	$= -x_1$
z_3	$\frac{4}{3}$	$-\frac{5}{3}$	$\left(\frac{40}{3}\right)$ (circled)	$\frac{1}{3}$	$= -y_3$
-1	$-\frac{1}{3}$	$-\frac{1}{3}$	$\frac{2}{3}$	$\frac{2}{3}$	$= f$
	$= z_2$	$= z_1$	$= u_3$	$= g$	

(c)

	y_2	y_1	y_3	-1	
u_2	$\frac{2}{15}$	$\frac{1}{6}$	$\frac{1}{10}$	$\frac{2}{5}$	$= -x_2$
u_1	$\frac{1}{6}$	$\frac{5}{24}$	$-\frac{1}{8}$	$\frac{1}{4}$	$= -x_1$
u_3	$\frac{1}{10}$	$-\frac{1}{8}$	$\frac{3}{40}$	$\frac{1}{20}$	$= -x_3$
-1	$-\frac{2}{5}$	$-\frac{1}{4}$	$-\frac{1}{20}$	$\frac{7}{10}$	$= f$
	$= z_2$	$= z_1$	$= z_3$	$= g$	

(d)

Figure: 5.64

The reader should check that these strategies do solve the game.

The examples just solved could have been worked directly, without the use of the Simplex Method. However, the Simplex Method works just as well for much larger games for which there is no easy direct method.

Exercises

1. If $q^0 = x^0/t$, where x^0 solves the maximum problem stated in (1), and $t = ex^0$, show that q^0 is an optimal strategy for player C in the matrix game G.
2. Solve the following games by the Simplex Method.

(a)

1	0	3
–2	3	0
–4	5	–6

(b)

–2	3	0	5	–6
3	–4	5	0	7
–4	5	–6	7	0

3. Consider the matching-pennies game with matrix

$G =$

1	–1
–1	1

   (a) Substitute it directly into the simplex format described in rule (5) above, and show that the Simplex Method breaks down.

   (b) Consider the Linear Programming problem denned in (1) with this G and show directly that it has no solution.

4. A passenger on a Mississippi riverboat was approached by a flashily dressed stranger (the gambler) who offered to play the following game: "You take a red ace and a black deuce and I'll take a red deuce and a black trey; we will simultaneously show one of the cards; if the colours don't match you pay me and if the colours match I'll pay you; moreover if you play the red ace we will exchange the difference of the numbers on the cards; but if you play the black deuce we will exchange the sum of the numbers. Since you will pay me with \$2 or \$4 if you lose and I will pay you either \$1 or \$5 if I lose, the game is obviously fair." Set up and solve the matrix game using the Simplex Method. Show that the game is not fair. Find the optimal strategies. [*Partial Ans:* The gambler will win an average of 25 cents per game.]
5. Consider the following game: R chooses 0 or 1 and reveals his choice to C; C chooses 0 or 1, but does not reveal his choice to

R; R then chooses 0 or 1 a second time. The sum of the three numbers chosen is computed and R receives odd sums while C receives even sums.

(a) Show that R has four strategies: 00, 01, 10, 11.

(b) Show that C has four strategies: (1) always choose 0, (2) choose the same as R, (3) choose opposite to R, (4) always choose 1.

(c) Show that the pay-off matrix is

	(1)	(2)	(3)	(4)
00	0	0	1	1
01	1	1	–2	–2
10	1	–2	1	–2
11	–2	3	–2	3

(d) Solve the game by the Simplex Method, finding its value and optimal strategies.

[*Ans.* $p^0 = \left(\frac{3}{4}, \frac{1}{4}, 0, 0\right), q^0 = \begin{pmatrix} \frac{3}{10} \\ \frac{9}{20} \\ \frac{1}{4} \\ 0 \end{pmatrix}, v = \frac{1}{4}.$]

6. Rework Exercise 5 assuming that the players choose 1 or 2 each time.

7. *The Silent Duel:* Two due lists each have a pistol that contains a single bullet and is equipped with a silencer. They advance towards each other in TV steps, and each may fire at his opponent at the end of each step. Neither knows whether his opponent has fired, and each has but one shot in the game. The probability that a player will hit his opponent if he fires after moving k steps is k/N. A player gets 1 if he kills his opponent without being killed himself, –1 if he gets killed without killing his opponent, and 0 otherwise. Each player has N strategies corresponding to firing after steps 1, 2,. . . , N. Let i be the strategy chosen by R and let y be the strategy chosen by C.

(a) If $i < j$, show that the expected pay-off to R is given by

$$\frac{N(i-j)+ij}{N^2}.$$

(b) If i > j, show that the expected pay-off to R is given by

$$\frac{N(i-j)-ij}{N^2}.$$

(c) If $i = j$, show that the expected pay-off to R is 0.

8. In Exercise 10, prove that the game is strictly determined and fair for $N = 2$, 3, and 4. Show that the optimal strategy for $N = 3$, 4 is to fire at the end of the second step in each case. For $N = 2$, show that any strategy is optimal.

9. In Exercise 10, show that the game is non-strictly determined and fair for $N = 5$, and find the optimal strategies.

   [*Ans.* $p^0 = (0, \frac{5}{11}, \frac{5}{11}, 0, \frac{1}{11}$ j), and q^0 is the column vector having the same components.]

10. A *symmetric matrix game* G is one for which $g_{ij} = -g_{ji}$ for i, j = 1, 2, ..., n. In other words, for every payment from C to R there is an equal payment from R to C. Show that every symmetric game is fair. [*Hint:* Show that if x^0 is optimal for R, then the column vector y with $y_k = x_k^0$ for $k = 1, \ldots, n$ is optimal for C. From this deduce that the value of the game is zero.]

11. Use Exercise 10 to show that the silent duel is fair for every N.

12. Consider a matrix game G with positive value in which the first row strictly dominates the second row. Show that in the simplex algorithm no entry in the second row will ever be chosen as a pivot in the first step.

13. Consider a matrix game G with positive value in which the first column dominates the second column. Show that if the pivot is chosen in the first column, after the end of the first simplex calculation the indicator for the second column will be non-negative.

Bibliography

Albers, Donald J. and Alexanderson, Gerald L.: *Mathematical People: Profiles and Interviews*, Birkhäuser, Boston, 1985.

Albers, Donald J., Alexanderon, Gerald L. and Reid, Constance: *More Mathematical People: Contemporary Conversations*, Academic Press, New York, 1994.

Anthony, Joby Milo: *In Eves Circle,* Math. Assoc. Amer., Washington, D. C., 1994.

Atwater, M. M.: *Multicultural Education: Inclusion for all*, University of Georgia, Athens, 1995.

Banks, J. A., and Banks, C.: *Handbook of Research on Multicultural Education*, MacMillan, New York, 1995.

Belenky, M. F., Clinchy, B. M., Goldberger, N. R., and Tarule, J. M.::*Women's Ways of Knowing*, Basic Books, New York, 1986.

Bell, M., and Burns, J.: *Counting and Numeration Capabilities of Primary School Children: A Preliminary Report*, University of Minnesota, Minneapolis, 1981.

Bellamy, T. and Wilcox, B.: *The Activities Catalog, An Alternative Curriculum for Youth and Adults With Severe Disabilities, Baltimore*, Paul H. Brookes Publishing Co., 1985.

Bender, M., and Valletutti, P. J.: *Teaching: A Curriculum Guide for Adolescents and Adults with Learning Problems*, Austin, 1982.

Birkhoff, Garrett,: *A Soucrebook in Classical Analysis*, Harvard University Press, Cambridge, 1973.

Bishop, A.: *Multicultural Literature for Children: Making Informed Choices,* Christopher-Gordon, Norwood, 1988.

Boyer, Carl: *A History of Mathematics,* John Wiley and Sons, New York, 1989.

Brolin, D.: *Life Centered Career Education: A Competency Based Approach,* Council for Exceptional Children, 1986.

Buhler, Walter: *Gauss: A Biographical Study,* Springer-Verlag, Berlin, 1981.

Burt, M.: *Black Inventors of America,* National Book Co., Portland, 1969.

Burton, David M.: *The History of Mathematics: An Introduction,* Wm. C. Brown Publishers, Dubuque, 1995.

Burton, L.: *Gender and Mathematics: An International Perspective,* Cassell Educational, London, 1990.

Canada, S. C.: *Science for Every Student: Educating Canadians for Tomorrow's World,* Ottawa, Canada, 1984.

Caporrimo, R.: *Gender, Confidence and Math: Why Aren't the Girls "Where the Boys Are"?* Psychological Association, Boston, 1990.

Cardano, J.: *The Book of My Life,* Dover, New York, 1962.

Carwell, H.: *Blacks in Science: Astrophysicist to Zoologist,* Exposition Press, New York, 1977.

Century, J. R.: *Making Sense of the Literature on Equity in Education,* Educational Development Center, Newton, 1994.

Clark, Ronald William: *The Life of Bertrand Russell,* Knopf, New York, 1976.

Cohen, M. R. and I. E. Drabkin: *A Source Book in Greek Science,* McGraw-Hill, New York, 1948.

Cole, M., and Griffin, P.: *Improving Science and Mathematics Education for Minorities and Women: Contextual Factors,* Wisconsin Center for Educational Research, Madison, 1987.

Council, N. R.: *Everybody Counts: A Report to the Nation on the Future of Mathematics Education,* National Academy Press, Washington, 1989.

Cuevas, G., and Driscoll, M.: *Reaching all Students with Mathematics,* National Council of Teachers of Mathematics, Reston, 1993.

Dauben, Joseph Warren: *Georg Cantor: His Mathematics and Philosophy of the Infinite,* Princeton University Press, New Jursey, 1990.

————: *The History of Mathematics from Antiquity to the Present: A Selective Bibliography,* Garland, New York, 1984.

Dedekind, Richard: *Essays on the Theory of Numbers,* Dover Publications, New York, 1954.

Descartes, Rene: *The Geometry,* Dover Publications, New York, 1954.

Dijksterhuis, E. J.: *Archimedes,* Princeton University Press, Princeton, 1987.

Duncan J. Melville: *Ration Computations at Fara: Multiplication or Repeated Addition?,* MacMillan, London, 2001.

Dunham, William: *Euler: The Master of Us All,* Math. Assoc. Amer., Washington, DC, 1999.

Eleanor, Robson: *Mesopotamian Mathematics, 2100-1600 BC: Technical Constants in Bureaucracy and Education,* Clarendon Press, Oxford, 1999.

Estrin, E. T.: *Alternative Assessment: Issues in Language, Culture and Equity,* Far West Laboratory, San Francisco, 1993.

Eves, Howard: *An Introduction to the History of Mathematics,* Saunders College Publishing, Philadelphia, 1988.

————: *Great Moments in Mathematics* Mathematical Association of America, Washington, 1981.

Fauvel, John and Jeremy Gray: *The History of Mathematics: A Reader,* MacMillan Press, London, 1987.

George, G. Hackman: *Sumerian and Akkadian Administrative Texts From Predynastic Times to the End of the Akkad Dynasty*, Yale University Press, Yale, 1958.

Gil, J. Stein: *Rethinking World-Systems: Diasporas, Colonies and Interaction in Uruk Mesopotamia*, University of Arizona Press, Arizona, 1999.

Gillespie, Charles Coulton: *Dictionary of Scientific Biography*, Charles Scribner's Sons, New York, 1980.

——————: *Pierre-Simon Laplace, 1749-1827: A Life in Exact Science*, Princeton University Press, Princeton, 1997.

Grattan, G. I.: *Companion Encyclopedia of the History and Philosophy of the Mathematical Sciences*, Routledge, London, 1994.

Guillermo Algaze: *The Uruk World System: The Dynamics of Expansion of Early Mesopotamian Civilization*, University of Chicago Press, Chicago, 1993.

Hall, T.: *Carl Friedrich Gauss, A Biography*, MIT Press, Cambridge, 1970.

Hans J. Nissen, Peter Damerow, Robert K. Englund: *Archaic Bookkeeping*, University of Chicago Press, Chicago, 1993.

——————: *The Early History of the Ancient Near East, 9000-2000 B.C.*, University of Chicago Press, Chicago, 1988.

Hollingdale, Stuart: *Makers of Mathematics*, Penguin, New York, 1991.

Hutchins, Robert Maynard: *Great Books of the Western World*, Encyclopedia Brittanica, Chicago, 1952.

Ignace J. Gelb, Piotr Steinkeller, Robert M. Whiting: *Earliest land Tenure Systems in the Near East: Ancient Kudurrus*, Oriental Institute of the University of Chicago, 1991.

Jens Hoyrup: *Remarkable Numbers' in Old Babylonian Mathematical Texts: A Note on the Psychology of Numbers*, Journal of Near Eastern Studies, 1993.

Joan, Fisher: *R. A. Fisher: The Life of a Scientist*, Wiley, New York, 1985.

Johnson, B., and Scharf, K.: *Menu Math for Beginners,* Remedia Publications, Scottsdale, 1982.

Kaluza, Roman: *Through a Reporter's Eyes: The Life of Stefan Banach,* Amer. Math. Soc., Washington, DC: 1996.

Keen, Linda: *The Legacy of Sonya Kovalevskaya: Proceddings of a Symposium,* Amer. Math. Soc., Prividence, 1987.

Kline, Morris: *Mathematical Thought from Ancient to Modern Times,* Oxford University Press, New York, 1972.

Koblitz, Ann Hibner: *A Convergence of Lives: Sofia Kovalevskaia, Scientist, Writer, Revolutionary,* Rutgers University Press, New Brunswick, 1993.

Kramer, E.: *The Main Stream of Mathematics: From the Earliest Beginning,* Oxford University Press, New York, 1951.

Leonardo, Pisano: *The Book of Squares,* Academic Press, New York, 1987.

Lucas, V. H. and Amey, M. J.: *Problem Solving Activities for Teaching Daily Living Skills – A Curriculum Handbook,* Cedars Press, Columbus, 1982.

Lutzen, Jesper: *Joseph Liouville 1809-1882: Master of Pure and Applied Mathematics,* Springer Verlag, New York, 1990.

Lwey, C. and Broad, Charlie D.: *Leibniz: An Introduction,* Cambridge University Press, London, 1975.

Mahoney, Michael Sean: *The Mathematical Career of Pierre de Fermat, 1601-1665,* Princeton University Press, Princeton, 1994.

Martin, Gardner: *Further Mathematical Diversions,* George Allen and Unwin, 1970.

May, Kenneth O.: *Bibliography and Research Manual for the History of Mathematics,* University of Toronto Press, Toronto, 1973.

Miller, L. S., and Glascoe, L. G.: *Life Centered Career Education: Activity Book One,* Council for Exceptional Children, Reston, 1986.

Mitchell, M. F., Belcher, L. A., and Brothers, N. S.: *Stepping Out with Language*, Communication Skill Builders, Tucson, 1988.

Monastyrsky, Michael: *Riemann, Topology and Physics,* Springer Verlag, New York, 1999.

Moore, M. H.: *Parent Partnership Training Program: A Comprehensive Skills Program for the Handicapped,* Walker Educational Book Corporation, New York, 1979.

Mordell, L. J.: *Reflections of a Mathematician,* Candian Mathematical Congress, Montreal, 1958.

Muir, Jane: *Of Men and Numbers: The Story of the Great Mathematicians,* Dover, New York, 1996.

Muller, Johann: *On Triangles,* University of Wisconsin Press, Madison, 1967.

Mullins, C.: *Life Skills Reading,* Educational Design, New York, 1980.

Newman, James R.: *The World of Mathematics,* Simon and Schuster, New York, 1956.

Nolan, Deborah: *Women in Mathematics: Scaling the Heights,* Math. Assoc. Amer., Washington, DC, 1997.

Ore, Oystein: *Cardona, the Gambling Scholar,* Dover, New York, 1965.

————————: *Niels Henrik Abel, Mathematician Extraordinary,* Chelsea, New York, 1957.

Pahre, P. and Husak, G.: *The Work Series. Senickley,* Hopewell Books, New York, 1976.

Peter, Damerow and Robert K. Englund, *The Proto-Elamite Texts from Tepe Yahya,* Cambridge, Mass., 1989.

Pitch, B.: *A Step-by-Step approach to Learning How to Fill Out Application Forms,* Frank E. Richards Publishing Co., New York, 1972.

Popovich, D., and Laham, S. L.: *The Adaptive Behavior Curriculum,* Paul H. Brookes Publishing Co., Baltimore, 1981.

Poundstone, William: *Prisoner's Dilemma: John Von Neumann, Game Theory and the Puzzle of the Bomb,* Anchor Books, 1993.

Radzik, March, K., and Strutchens, M.: *Multicultural Education: Inclusion for All,* The Universiy of Georgia, College of Education, 1994.

Reid, Constance: *Courant in Gottingen and New York: The Story of an Improbable Mathematician,* Copernicus Press, 1996.

————————: *Julia: A Life in Mathematics,* Math. Assoc. Amer., Washington, DC, 1996.

————————: *Neyman: From Life,* Springer Verlag, New York, 1982.

————————: *The Search for E.T. Bell: Also Known as John Taine,* Math. Assoc. Amer., Washington, DC, 1993.

Richey, J.: *Survival Vocabularies: Banking, Clothing, Credit, Driver's License, Drugstore, Entertainment, Job application, Medical, Restaurant, and Supermarket Language,* Janus Books, Hayward, 1979.

Roger, Cooke: *The Mathematics of Sonya Kovalevskaya,* Springer Verlag, New York, 1984.

Ronald, Calinger: *Classics in Mathematics,* Prentice Hall, New Jursey 1995

Rouse, Ball and Coxeter, H.: *Mathematical Recreations and Essays,* University of Toronto Press, Toronto, 1974.

Rowe, David and John McCleary: *The History of Modern Mathematics,* Academic Press, New York, 1989.

Scharf, K., and Johns, B.: *Market Math,* Remedia Publications, Scottsdale, 1981.

Schechter, Bruce: *My Brain is Open: The Mathematical Journeys of Paul Erdos,* Simon and Schuster, New York, 1998.

Schwartz, Laurent: *Un Mathematicien Aux Prises Avec le Siècle.* Odile Jacob, Paris, 1997.

Simmons, George Finlay: *Calculus Gems: Brief Lives and Memorable Mathematics,* McGraw-Hill, New York, 1992.

Smith, David Eugene: *A Source Book in Mathematics,* Dover Publications, New York, 1959.

Somers, D. J.: *Learning Functional Words and Phrases,* Frank E. Richards Publishing, New York, 1977.

Stein, Sherman: *Archimedes: What Did He Do Besides Cry Eureka?* Amer. Math. Soc., Washington, DC, 1999.

Stillwell, J.: *Mathematics and Its History,* Springer Verlag, New York, 1989.

Struik, Dirk J.: *A Concise History of Mathematics,* Dover Publications, New York, 1999.

————————: *A Source Book in Mathematics, 1200—1800,* Princeton University Press, Princeton, 1986.

Tarski, A. and Robinson, R. M.: *Undecidable Theories,* Amsterdam, 1953.

Tarski, A.: *Logic, Semantics, Metamathematics,* Oxford, New York, 1956.

Taylor, Harold and Taylor, Loretta: *George Pólya: Master of Discovery,* Math. Assoc. Amer., Washington, DC, 1993.

Turnbull, Herbert Westren: *The Great Mathematicians,* New York University Press, New York, 1961.

Udolph, L. K., and Sherman, E.: *Shop Talk: A Prevocational Language Program for Retarded Students,* Research Press, Champaign, 1983.

Ulam, Stanislaw M.: *Adventures of a Mathematician,* Scribner, New York, 1976.

Valletutti, P. J. and Bender, R.: *Teaching the Moderately and Severely Handicapped,* Communication and Socialization, Austin, 1985.

Valletutti, P. J., Bender, M. and Bender, R.: *Teaching the Moderately and Severely Handicapped,* Motor Skills and Household Management, Austin, 1985.

Valletutti, Peter J. and Bender, M.: *A Functional Curriculum for Teaching Students with Disaiblities,* Austin, 1996.

Wehman, P., and McLaughlin, P. J.: *Vocational Curriculum for Develomentally Disabled Person,* Austin, 1980.

Wehman, P., Renzagliz, A. and Bates, P.: *Funcational Living Skills for Moderately and Severely Handicapped Individuals,* Pro-Ed, Austin, 1985.

Whitrow, Magda: *ISIS Cumulative Bibliography: A Bibliography of the History of Science Formed From ISIS Critical Bibliographies,* Mansell, London, 1971.

Index

G

I

L

M

N

O

P

R

❑❑❑